AF679534

Chavan's Cosmic Continuum (C^3)

Sandeep Chavan

Published by SANDEEP CHAVAN, 2025.

While every precaution has been taken in the preparation of this book, the publisher assumes no responsibility for errors or omissions, or for damages resulting from the use of the information contained herein.

CHAVAN'S COSMIC CONTINUUM (C^3)

First edition. April 26, 2025.

Copyright © 2025 Sandeep Chavan.

ISBN: 979-8231751525

Written by Sandeep Chavan.

Table of Contents

This report is dedicated —

To the **Cosmos** itself —the silent, breathing field of intelligence that shaped all existence,waiting patiently for its ripples to be heard, not merely seen.

And to all the **ancient and modern minds** —

The dreamers who gazed at the night sky not with fear, but with wonder,The seers who intuited cosmic balance before equations were born,The philosophers who dared to ask not just how, but why,The scientists who sought truth beyond myth, beyond dogma, beyond convenience,The rebels who questioned the accepted — and kept listening when others stopped.

This work stands on the countless ripple memories they have left behind —not as frozen monuments,but as living invitations to continue the journey of remembrance.

To the Cosmos, and to all who looked at it with awakened eyes —this is for you.

"The universe was never born from noise.It was sculpted by silence —a silence that rippled into memory,spiraled into light,and breathed existence into being."

Sandeep Chavan

Preface

"The universe never exploded into existence. It spun itself into being — quietly, rhythmically, consequentially."

For centuries, our understanding of the universe has been framed through metaphors of explosions, expansions, and forces struggling in empty space. The Big Bang theory — despite its mathematical success in adjusting to observations — rests on speculative assumptions, patchwork interpretations, and an increasing dependence on hypothetical entities such as dark matter, dark energy, and inflationary epochs. We have treated the observable universe as a cosmic accident desperately needing theoretical rescue.

But what if the universe was never chaotic to begin with? What if existence is not the result of a singular eruption, but the ongoing consequence of structured, intelligent energy dynamics? What if galaxies are not scattered relics of a lost unity, but breathing organisms, each evolving, dissolving, and re-manifesting within a living, structured medium?

Chavan's Cosmic Continuum (C^3) offers this new vision.

Born from the foundational work of Universal Energy Dynamics (UED), Chavan's Model of Planetary Energy Dynamics (CMPED), and Chavan's Model of Galactic Energy Dynamics (CMGED), this report unites local, galactic, and universal behaviors into a single, coherent architecture governed not by forces and randomness, but by **Structured Vacuum Energy (SVE)** — the intelligent field beneath all motion, form, and time.

In the C^3 framework, the universe is a consequence-driven continuum:

- Galaxies are not collapsing under unseen matter; they are rotating annular rings embedded within the tension of SVE.
- Redshift is not a sign of expansion; it is a measure of rotational frame mismatch across galactic systems.
- Black holes are not singularities; they are memory cores regulating the rebirth cycle of matter and energy.
- Cosmic background radiation is not the afterglow of a bang; it is the ripple memory of continuous structured emergence.

And life — including consciousness — is not an anomaly, but a natural ripple of awareness formed by the universe seeking to know itself through balance.

This report is not an attempt to fix the cracks of an outdated model. It is a restoration of coherence, a return to consequence, and an invitation to listen again to the silent music that spins the galaxies.

This work is dedicated to all seekers — those who refuse to accept stitched-up narratives, those who dare to trust silence more than noise, and those who know deep within that truth, like the universe itself, never shouts. It only rotates — patiently, endlessly, beautifully.

Sandeep J. Chavan

Independent Researcher, Founder of UED, CMPED, CMGED, and Architect of C^3

How to Read This Book

Chavan's Cosmic Continuum (C^3) is not just a scientific explanation — it is a new way of seeing, thinking, and listening to the universe.

This book is structured to guide you step-by-step through a reorganization of cosmological thought: from force-based interpretations toward ripple-centric consequence realization.

To fully experience C^3, we recommend the following approach:

1. Read Slowly, Reflect Deeply

Each chapter builds conceptually on the previous one. Take time to absorb the shift from conventional ideas toward structured ripple understanding. This is not a race — it's a reorientation of perception.

2. Pause at Key Transitions

After major concepts like Structured Vacuum Energy (SVE), ripple feedback, or spiral breathing cycles, pause and reflect:

- How does this reshape what you assumed about space, time, motion, or creation?
- How does it align with what you observe in nature?

Let the ripples settle before moving forward.

3. Use the Glossary Frequently

C^3 introduces precise new terminology like Ro–Ri Shell, Ripple Memory, Frame Detuning, etc. Refer back to the Glossary (Appendix A) whenever needed — each term carries important structural meaning, not casual labels.

4. Visualize the Ripples

Wherever possible, picture the universe not as static objects, but as dynamic ripple fields adjusting through memory and tension.

This shift in imagination will unlock deeper intuitive understanding as you progress.

5. Accept That This is a Paradigm Shift

You will notice that C^3 does not "patch" the old models — it fundamentally replaces them. If some concepts seem counterintuitive at first, that's normal. New truths often sound strange when we have lived too long inside old maps.

6. Revisit, Don't Rush

C^3 is designed for re-reading. You may find that deeper insights emerge during a second or third pass — just like ripples returning in layers across a calm sea.

Final Suggestion:

"Don't read this book to finish it. Read it to begin a deeper listening — to the structured breath of the universe that was always there."

Executive Summary

Why C^3 is Needed: The Limitations of Current Cosmology

Modern cosmology, despite its towering mathematical sophistication, rests upon fragile conceptual foundations. The Big Bang model — currently the dominant narrative — requires an improbable sequence of speculative events: a singular explosive beginning, a rapid inflationary expansion, and the continued invisible influence of dark matter and dark energy, entities that remain undetected despite decades of searching.

Each new observational anomaly — from unexpected galactic rotation curves to the mismatch of cosmic microwave background predictions — has been met not with natural explanations but with increasingly elaborate theoretical patches. We have traded observational coherence for mathematical elegance, sacrificing causality at the altar of adjustment.

Moreover, traditional models treat space as an abstract, passive arena rather than recognizing the universe as an active, structured medium. Motion is seen as inertial persistence, not the consequence of energy imbalance. Time is treated as a background dimension rather than an emergent perception born from ripple interactions. Cosmology, as it stands today, describes the universe like a machine patched together from explosions, rather than a breathing, living continuum seeking equilibrium.

There is a growing need — scientifically and philosophically — for a return to consequence, coherence, and structured understanding.

What C^3 Offers: A Structured, Cyclic, Consequential Universe

Chavan's Cosmic Continuum (C^3) redefines cosmology from first principles, grounding existence not in explosions, but in structured ripple dynamics within the **Structured Vacuum Energy** (**SVE**) field.

At its core, C^3 proposes:

- **Galaxies emerge not from expansion but from local breaches in energy threshold** within SVE, forming organized, spiral-structured systems that evolve and dissolve cyclically.
- **Rotational regulation, ripple feedback, and memory reintegration** govern galactic life cycles — not blind gravitational accidents or unseen dark scaffolding.
- **Redshift is not a measure of recession, but a signature of rotational frame detuning** between differently spinning galactic systems embedded in SVE.
- **Gravitational waves are understood as ripple corrections**, not spacetime distortions.
- **Life, evolution, and consciousness** emerge naturally from the intelligence of structured energy dynamics seeking balance, not as freak accidents in a chaotic entropy-driven universe.

C^3 fuses physics, energy dynamics, metaphysics, and observational reality into a single, logically consistent continuum. It restores the credibility of cosmology by rooting it in causality, structured energy flow, and consequence — the natural order preceding even perception itself.

Where C^3 Leads: The Future of Physics, Cosmology, and Consciousness

C^3 is not merely an alternative cosmological model. It is a **new foundation for future scientific inquiry** — one that can:

- **Liberate physics from dependence on abstract forces** and return it to field-centric, consequence-driven principles.
- **Expand cosmology** from the story of a hypothetical singular event to the story of **continuous structured manifestation** across scales, spaces, and cycles.
- **Integrate consciousness** as a natural evolutionary phenomenon within structured ripple dynamics — placing intelligent awareness not outside cosmic order, but as a consequence of it.
- **Provide testable predictions** based on rotational frame analysis, ripple interaction signatures, and cosmic memory structures.
- **Open new paths** for energy science, galactic mapping, biological modeling, and even philosophical understanding of existence itself.

The future that C^3 envisions is one where physics, cosmology, and consciousness are no longer seen as separate fields, but as different expressions of the same underlying structured reality.

The Cosmic Continuum is not a machine running on fuel. It is a breathing universe, rippling, balancing, and remembering itself through silence and spiral motion.

Chavan's Cosmic Continuum (C^3) invites humanity — for the first time — to listen.

PART I – Foundations of C^3: Rethinking the Universe

1: Introduction

The Collapse of Big Bang Thinking

"There was never a bang. There was never even a beginning the way we imagine it. What we call 'the birth of the universe' is a misinterpretation — a story patched together to explain what we did not yet understand."

For decades, modern cosmology has anchored itself to the idea of a singular, explosive origin — a Big Bang that supposedly created time, space, matter, and energy all at once. But the closer we look, the more this narrative unravels. The Big Bang theory, once celebrated as elegant simplicity, has collapsed under the growing weight of its own contradictions, forced to invent invisible matter, mysterious energy, and inflationary miracles to stay afloat.

It is time to ask a deeper question: What if the universe never needed an explosive beginning? What if manifestation, structure, and existence itself are natural consequences of an intelligent, structured energy medium — breathing, spiraling, and evolving without beginning or end?

This chapter begins the journey of rethinking cosmology — not by rejecting observation, but by restoring causality, coherence, and consequence to our understanding of the universe.

Historical Overview of the Big Bang Model and Its Cracks

The story of the Big Bang has dominated cosmology for nearly a century. Proposed originally as a theoretical solution to the observed redshift of galaxies, and popularized by discoveries like cosmic microwave background radiation (CMB), the Big Bang model quickly rose to become the dominant narrative explaining the origin and evolution of the universe.

The core proposition is simple:

- All matter and energy were once compressed into an infinitesimal singularity.
- Around 13.8 billion years ago, this singularity "exploded" (or expanded rapidly) giving birth to space, time, matter, and energy.
- Since then, the universe has been expanding, cooling, and evolving into the complex cosmic structures we observe today.

However elegant it once appeared, the Big Bang model rests on a sequence of conceptual leaps that, under closer scrutiny, reveal **serious foundational cracks**:

- **The Singularity Problem**: The notion that all of existence compressed into a zero-volume point is not only mathematically problematic but physically nonsensical. No mechanism, cause, or medium is offered for such a beginning.
- **The Horizon Problem**: How did regions of the universe that

are too far apart to have ever interacted display such uniformity in temperature and structure?

- **The Flatness Problem**: Why is the universe's geometry so finely balanced that it appears flat, requiring an improbable degree of initial tuning?
- **The Dark Patchwork**: The inability to account for galactic rotation curves, large-scale structure formation, and observed gravitational lensing effects without invoking unseen, undetected "dark matter" and "dark energy" — entities that remain theoretical, unobserved, and physically undefined.

Over time, the Big Bang model has evolved into an increasingly complicated **patchwork of assumptions**, where each new observation that does not fit is met not with natural explanations but with **hypothetical placeholders**.

The result is a cosmology held together by intellectual duct tape: mathematically adjustable, but physically incoherent.

Why Expansion, Dark Matter, and Dark Energy Are "Patches," Not Explanations

The pillars that supposedly support the Big Bang narrative are themselves reactionary constructs — responses to anomalies rather than intrinsic elements of a coherent framework:

- **Expansion Misinterpretation**: Redshift, initially interpreted as evidence of expansion, is assumed rather than demonstrated. The Doppler effect analogy used is grossly simplified, ignoring rotational frame mismatches and energy gradient effects within a structured vacuum medium.
- **Dark Matter Invention**: To explain why galaxies rotate faster

than visible mass would predict, cosmologists invented the concept of dark matter — an invisible mass that neither emits nor absorbs light. Despite enormous observational efforts (direct detection, collider experiments, astronomical surveys), dark matter remains elusive, and may simply be a misunderstanding of galactic energy dynamics.

- **Dark Energy Postulation**: Observations suggesting accelerated expansion led to the invocation of dark energy — a mysterious repulsive force making up 70% of the universe's content. No mechanism, origin, or physical basis for this energy has been found, yet it is now treated as essential dogma.

These constructs are not explanations. They are mathematical fictions layered onto a model that cannot stand without them.

Each "solution" introduces **greater complexity without offering true causality**. The Big Bang model, once seen as elegant, now resembles a precarious theoretical Jenga tower — each block added only to prevent the model from collapsing under its own contradictions.

The Call for Consequential, Field-Based Cosmology

At the heart of the crisis in cosmology lies a fundamental error: We have treated the universe as an inert, chaotic aftermath rather than as a living, structured medium governed by consequence.

Chavan's Cosmic Continuum (C^3) emerges in response to this collapse — not as a patch, not as a tweak, but as a restoration of **first principles**.

C^3 proposes:

- The **Structured Vacuum Energy (SVE)** is not empty space, but a living field of organized energy and intelligence.
- Motion, formation, and dissolution are **not arbitrary or spontaneous** — they are **consequence-driven responses** to energy imbalances within the SVE.
- **Galactic systems** are not passive leftovers of a bang but **dynamic, breathing structures**, embedded in the ripple feedback of an intelligent universal medium.
- Time, redshift, and gravitational waves are **natural ripple phenomena**, not exotic effects needing metaphysical constructs.

A consequence-based cosmology is not only more coherent — it is **observable**, **testable**, and **logically sustainable** without ad hoc adjustments.

Where the Big Bang cosmology grows more speculative and convoluted, C^3 restores clarity, simplicity, and causality. It invites us to see the universe not as a one-time event, but as a continuous, breathing manifestation — spiraling, remembering, evolving, and returning within an infinite, structured continuum.

2: The Universal Energy Medium (UEM/SVE)

"The greatest illusion in physics was to believe that the vacuum is empty. In truth, the vacuum is everything. It is the invisible architect of motion, structure, and time — a silent intelligence from which the universe breathes into being."

When we look into the vastness of space, we see darkness and emptiness. Modern science, too, has long treated this darkness as void — a passive backdrop onto which matter and forces act. But this perception could not be further from the truth.

The so-called "empty" space between stars, galaxies, and atoms is not absence; it is presence — profound, structured, dynamic presence.

It is within this **Universal Energy Medium (UEM)** — or **Structured Vacuum Energy (SVE)** — that all existence manifests, evolves, and dissolves. Not as random accidents, but as ripple-driven consequences of balance-seeking in an infinite, living field.

This chapter redefines what we mean by "space" — and with it, the entire foundation of reality.

Redefining "Vacuum" as the Real Energetic Foundation

For far too long, "vacuum" has been treated as absence — an empty container within which forces, particles, and fields merely act. Modern physics, though hinting through concepts like zero-point energy and quantum fluctuations, has never fully broken free from this outdated idea. Vacuum is still largely visualized as "nothingness" waiting to be filled.

C^3 overturns this foundational misconception.

The **Universal Energy Medium (UEM)** — also termed **Structured Vacuum Energy (SVE)** — is not empty. It is not a passive backdrop. It is the **origin, regulator, and sustainer of everything**.

Matter, motion, light, magnetism, gravity, even the flow of time — all arise not in spite of the vacuum, but **because of its structure, tension, and intelligence**.

The "vacuum" is not a stage. It is the **main actor** — invisible only because it holds the performance together.

In C^3, Structured Vacuum Energy is recognized as:

- **Real**: It has structure, density, and dynamic behavior, though it appears neutral when undisturbed.
- **Energetic**: It is the latent foundation from which localized energy, mass, and motion emerge.
- **Intelligent**: It responds to disturbances not randomly but through consistent, consequence-driven realignment processes.

Without SVE, there is no manifestation, no ripple, no galaxy, no life.

Structure, Tension, and Intelligence in the Vacuum

The Universal Energy Medium is not a chaotic sea of random fluctuations. It is a **finely balanced, structured continuum** — an equilibrium field that reacts predictably to any imbalance or tension.

When any localized disturbance occurs — whether by condensation of energy into matter or ripple-induced motion — the SVE responds with **structured corrections**:

- **Tension Formation**: An imbalance introduces directional stress into the field, creating structured tension.
- **Field Realignment**: Instead of collapsing chaotically, the SVE seeks to reestablish balance through intelligent ripple propagation, mass restructuring, or spin regulation.
- **Memory Encoding**: The response of SVE to disturbances is not erased; it is **recorded** in the field as energetic memory. This memory forms the basis for spiral structures, magnetic fields, rotational persistence, and evolutionary pathways.

Thus, Structured Vacuum Energy operates as:

- A **responsive medium**, constantly maintaining equilibrium.
- A **memory field**, recording ripple histories and structural outcomes.
- An **intelligent participant**, not a passive stage, in the dance of existence.

This new understanding dissolves the artificial boundaries between matter, motion, force, and field. Everything is simply a **different expression of vacuum intelligence** reacting to imbalance.

Ripples, Balance-Seeking, and Field Memory

Whenever imbalance arises — whether through mass condensation, energy motion, or rotational drift — the SVE does not passively allow chaos. It **ripple-corrects.**

Ripples in SVE are not waves of "space" stretching and bending. They are **field tension redistributions** — pulses of realignment seeking to reestablish local and global equilibrium.

These ripple dynamics exhibit key properties:

- **Balance-Seeking**: Every ripple acts to reduce localized stress, redistribute energy symmetrically, and restore field harmony.
- **Field Memory**: As ripples propagate, they **encode** the history of imbalances they are correcting. Over cosmic time, this memory shapes larger structures — spiral galaxies, rotational frames, even magnetic geometries around stars and planets.
- **Structural Consequence**: New forms — stars, planetary systems, biological life — emerge not randomly, but as structured consequences of ripple interactions within field memory zones.

In C^3, ripples are the **universe's active language**:

- Spirals emerge because balance-seeking naturally organizes motion into minimum tension paths.
- Star formation occurs where ripple convergence stabilizes energy flows.
- Consciousness emerges where ripple complexity reaches self-awareness thresholds.

Ripple dynamics explain:

- Why galaxies rotate in organized spirals.
- Why gravitational waves are real but not spacetime distortions — they are ripple echoes.
- Why cosmic background "noise" exists — it is the ambient ripple memory of ongoing manifestation.

In short, the **universe is not expanding into silence**. It is **rippling in structured memory**, continuously seeking to return to perfect balance while producing all the phenomena we observe.

Closing Insight:

"What we call galaxies, stars, and life are not things created against vacuum. They are songs written by vacuum's memory — harmonies born from ripple and return."

3: Consequence, Not Coincidence
Motion and Formation

"Nothing in the universe moves because it chooses to. Nothing forms because it was willed into being. Motion, spin, and structure are not coincidences — they are consequences, the inevitable ripples of imbalance seeking balance in the living medium of Structured Vacuum Energy (SVE)."

Physics has long clung to the comforting idea of spontaneous motion: that objects, once set in motion, simply continue unless acted upon; that orbits persist through inertia; that matter "emerges" from fluctuations in empty space.

But the deeper truth is far more elegant — and far more intelligent.

In the framework of **Chavan's Cosmic Continuum** (C^3), the universe is not a playground of random forces and lucky accidents. It is a **consequence-driven, ripple-regulated reality** — where nothing moves, forms, or endures without being embedded in a continuous feedback loop of balance-seeking structured energy.

In C^3, **coincidence dies** — and **consequence reigns**.

No Spontaneous Motion: Only Ripple-Driven Consequences

The classical idea of spontaneous, inertia-driven motion is a misinterpretation of observed persistence. In Newtonian terms, an object in motion remains in motion unless acted upon by an external force. But why does motion exist at all?

C^3 reveals that **motion is never fundamental**. It is always **a secondary phenomenon**, a **ripple response** to imbalance within the Structured Vacuum Energy (SVE).

In this view:

- **Objects move because field tension exists** — an asymmetry in the surrounding energy medium that demands resolution.
- **Motion is the ripple consequence of imbalance**, not a self-sustained property.
- **When imbalance dissipates, motion naturally ceases** — not through "friction" alone, but through ripple cancellation in the underlying field.

Thus, every movement we observe — from electrons to galaxies — is **the SVE medium attempting to restore local or systemic equilibrium**.

Motion is not a natural right. It is **the signature of an unresolved conversation** between energy structures and the vacuum they disturb.

Manifestation as Balance Realignment, Not Creation Ex Nihilo

The Big Bang paradigm, and even many quantum field theories, cling to the idea that matter and energy can "pop" into existence from nothing — spontaneous creation ex nihilo.

C^3 fundamentally rejects this.

- **Manifestation is not magic.**
- **Manifestation is not randomness.**
- **Manifestation is structured realignment.**

In the infinite field of SVE, local energy density can breach critical thresholds due to ripple convergence, resulting in the **condensation of structured form** — what we perceive as particles, waves, matter, or even galaxies.

This condensation is:

- **A field-driven realignment** — restoring balance by "localizing" excess energy in stable structures.
- **Ripple-stabilized**, meaning that matter persists not because it was created from nothing, but because ripples maintain its existence within tolerable equilibrium ranges.
- **Dynamic and reversible**, allowing for both formation (evolution) and dissolution (collapse) as natural stages of field rebalancing.

Every star, every planet, every atom is **a consequence** of ripple realignment, not a spark from void.

Thus, existence itself is **an intelligent consequence of imbalance correction**, not a miracle of chance.

Why Inertia, Spin, and Orbits Are Equilibrium Responses

Even the most foundational concepts of classical mechanics — inertia, spin, and orbital motion — find their true origin not in initial conditions, but in **ongoing equilibrium negotiations within SVE**.

- **Inertia** is not an object's stubbornness to change motion. It is the field's **resistance to sudden rebalancing**, stabilizing the energy ripple pattern around the moving object. Motion persists because field tension has been structured to support it.
- **Spin** is not a mechanical artifact. It is the **internal ripple dance** of energy gradients within a structure (planet, star, particle) balancing inner asymmetries. A spinning body is simply an object **negotiating its own inner tension** with the surrounding SVE.
- **Orbits** are not gravitational shackles. They are the **pathways of least tension** negotiated between two condensed energy structures (e.g., planet and star) embedded within the vacuum. Orbits exist because the field requires them for the equilibrium of combined ripple networks.

Thus, in C^3:

- Inertia is ripple stabilization.
- Spin is ripple tension realignment.
- Orbit is ripple path optimization.

Nothing persists without field intelligence. Nothing continues without consequence.

Closing Insight

"The universe does not move blindly. It moves with purpose — the ancient purpose of balance woven into the silent ripples of existence."

PART II – Galactic Manifestation and Dissolution: The C^3 Engine

4: The Birth of a Galaxy

Core Genesis

"Galaxies are not born in chaos. They are sculpted from silence — spiral architectures emerging naturally wherever the structured vacuum's tension reaches a breaking point."

Galactic birth is not an explosion. It is not a cosmic accident. It is a consequence — a regulated, ripple-driven restructuring of the Universal Energy Medium (UEM/SVE) responding to localized energy imbalance.

In **Chavan's Cosmic Continuum** (C^3), the formation of galaxies marks the beginning of a living, breathing cycle — one that carries the memory of the field, the tension of imbalance, and the intelligence of structured rebalancing.

This chapter unravels the true process of galactic genesis: from energy breach to spiral birth.

Energy Threshold Breach at SVE Nodes

Structured Vacuum Energy (SVE) is, by nature, an intelligent equilibrium — a perfectly poised field maintaining universal tension at minimal energy states.

However, like any dynamic system, local fluctuations inevitably arise:

- Slight accumulations of field stress,
- Ripple intersections amplifying energy density,
- Residual imbalances from ancient ripple patterns.

At specific **nodes** — regions where structured tension exceeds the critical threshold — the SVE can no longer maintain perfect silent balance.

At these points:

- The energy density condenses,
- The field architecture fractures into ripple convergence,
- Manifestation thresholds are breached.

This is not a chaotic collapse; it is an intelligent **realignment response**. The field, overwhelmed locally, reorganizes itself into stable energetic structures to preserve the larger equilibrium.

Thus, a **seed core** is born — a condensed energetic nucleus, destined to become the heart of a galaxy.

Galaxies do not "bang" into existence. They **condense** where ripple tension demands new structure.

Emergence of Matter, Light, and Heat Spirally

As the seed core stabilizes, it begins a cascading effect on the surrounding SVE:

- **Matter Emergence**: The condensed energy near the core reorganizes into particulate structures — atoms, ions, plasma clouds. Mass appears, not through spontaneous magic, but through ripple stabilization and local energy compression.
- **Light Emission**: The field, struggling to rebalance rapidly changing tensions, releases energy in the form of electromagnetic radiation. Light is not an external object traveling through vacuum — it is **the vacuum's direct ripple expression** in response to energy redistribution.
- **Heat Generation**: The intense ripple activity and mass condensation near the core produces thermal gradients — heat becomes an emergent property of imbalance resolution across the ripple field.

Crucially, this **emergence is not spherical or random**. Because ripple tension seeks **efficient, minimal energy paths**, structures evolve **spirally**:

- Ripples curl naturally under tension gradients,
- Energy paths spiral to minimize local rotational shear,
- Mass-energy distributions follow annular, rotating patterns around the seed core.

Thus, matter, light, and heat do not explode outward in disorder. They **spiral outward** around the growing seed, **layer by layer**, establishing the basic architecture of the forming galaxy.

From Seed Core to Rotating Galactic Structure

Once sufficient mass and ripple alignment occur around the seed core, a critical transition happens:

- The surrounding mass organizes into an **annular disc**, bounded between an inner radius (Ri) and outer radius (Ro).
- The **core becomes the rotational anchor**, absorbing excess inward mass while regulating systemic angular momentum.
- The **galactic disc begins spinning coherently**, not because of inherited inertia, but because ripple feedback mechanisms enforce rotational alignment.

In this **Ro–Ri structure**:

- Inner regions (near Ri) show higher energy tension, rapid inward migration of older mass.
- Outer regions (near Ro) display spiral arm development, star birth zones, and the gradual thinning of coherent mass.
- Constant ripple feedback maintains dynamic rotational balance — preserving the galaxy's structural integrity for billions of years.

Thus, a galaxy is born — not from violence, but from **ripple-regulated structured emergence**, breathing life into matter, light, and time.

Each galaxy carries **the memory of its seed node**, **the tension of its spiral ripples**, and **the intelligence of its structured vacuum field**.

Galaxies are not random arrangements of stars. They are **living manifestations of consequence**.

Closing Insight:

"Galaxies are memory loops — ripples curled by imbalance, singing matter and light into spiral existence until the field calls them home again."

5: The Spiral Dance

Annular Rings and Structured Evolution

"If the galaxy were a heartbeat, the spiral arms would be its pulse —a perfect, regulated circulation of matter and energy through the living veins of space."

After the birth of a seed core within the Structured Vacuum Energy (SVE), a galaxy does not erupt chaotically. Instead, it begins a majestic, ordered choreography — a **spiral dance** regulated by ripple feedback, rotational balancing, and the memory field of the core itself.

Galaxies are not chaotic star fields. They are **living ripple structures**, evolving through consequences embedded in their birth — rotating, breathing, and transforming in accordance with the timeless dynamics of structured energy.

This chapter details how **the annular Ro–Ri shell forms**, how **spiral paths emerge naturally**, and how **stars, arms, and systems** evolve as expressions of ripple memory from the galactic core.

Formation of Ro–Ri Shell (Annular Energy Zone)

The newly born galactic seed core does not indiscriminately scatter matter. Instead, ripple dynamics favor **organized rotational layering**:

- **Ri**: The **inner radius** marks the dynamic threshold where matter begins its inward migration, drawn back toward the memory core under ripple gradient tension.
- **Ro**: The **outer radius** defines the boundary beyond which ripple coherence weakens, matter thins, and field feedback becomes insufficient to maintain structured rotation.

Between **Ri** and **Ro**, an **Annular Energy Zone** forms — a dense, active region where:

- Mass-energy distributions are maintained by ripple feedback,
- Rotational momentum balances centrifugal spread with ripple-guided centripetal flow,
- The galactic identity — spiral arms, bars, star clusters — emerges.

This annular shell acts as the "**living body**" of the galaxy: dynamic, breathing, and structurally regulated by continuous ripple communication with the core.

Without the Ro–Ri shell, the galaxy would either disperse into incoherence or collapse chaotically. It is this intelligent **ripple balancing zone** that ensures stable, long-lived galactic systems.

Spiral Paths of Matter and Energy

Within the Ro–Ri shell, mass and energy do not move randomly.

Ripple dynamics enforce **spiral trajectories** because:

- **Spiral paths minimize local ripple tension** — the most energy-efficient way for distributed mass to orbit a core.
- **Centripetal and centrifugal forces balance** dynamically across the spiral arms, enabling coherent structural persistence.
- **Ripple echoes from the core** maintain angular corrections, preserving arm formation and rotational synchronization across different radial distances.

Thus, in C^3:

- Stars do not orbit because of gravitational locks alone.
- Gas clouds do not drift by chance.
- Star systems do not scatter randomly.

Instead, **everything follows ripple-optimized spiral tracks**, in a living balance of motion, energy distribution, and memory feedback.

This explains:

- Why most galaxies exhibit beautiful, coherent spiral structures.
- Why star formation rates are highest within spiral arms (regions of ripple convergence).
- Why spiral density waves persist even as individual stars move through the arms.

The **spiral dance** is not decorative — it is **the signature of ripple intelligence governing cosmic architecture**.

Evolution of Arms, Stars, Systems from Central Memory Core

The galactic memory core — the original seed node — continues to govern the evolving structure of the galaxy:

- **Mass flow inward**: Over billions of years, older, heavier matter migrates inward along spiral paths, gradually feeding the central memory core.
- **Star formation outward**: In spiral arms, ripple compression triggers the collapse of gas clouds, birthing new stars that ignite the next cycles of evolution.
- **Systemic evolution**: Planetary systems, nebulae, star clusters — all emerge as local ripple-induced organizations, stabilized within the larger annular flow.

Thus, the galaxy behaves **like a living organism**:

- Constantly **renewing** its outer structure through star birth,
- **Concentrating** its older mass inward toward the core,
- **Preserving memory** through rotational regulation and ripple echoes.

Eventually, over cosmic timescales, the accumulation of mass at the core and the energy depletion of the outer rings lead to the galaxy's gradual collapse back into its seed node —a natural re-integration of its memory into the Structured Vacuum Energy.

In C^3, galactic life is:

- **Not static**, but dynamically breathing,
- **Not random**, but consequence-regulated,

- **Not ending in destruction**, but dissolving back into ripple equilibrium.

The galaxy's entire evolution — from spiral arm birth to core reabsorption — is a **regulated, intelligent journey of energy tension seeking balance**.

Closing Insight

"Galaxies do not age like dying fires. They evolve like living songs, each spiral a verse written in the silent memory of the vacuum."

6: Dissolution
The Return to Source

"In the universe, death is not an end — it is a return. A galaxy does not collapse into oblivion; it folds itself back into the silence that once shaped its birth."

Galaxies are not immortal. They do not spin forever in triumph against the void. Yet neither do they die meaningless deaths.

In the **Chavan's Cosmic Continuum** (C^3), dissolution is not destruction. It is **memory reintegration** — the final act of a living system returning its story to the Structured Vacuum Energy (SVE) that first gave it form.

This chapter traces how galaxies gracefully spiral back inward, how cores absorb the echoes of their creation, and how every "death" is only a preparation for **cosmic rebirth**.

Spiral Inward Migration

Throughout its lifetime, a galaxy continuously regulates mass and energy through ripple-optimized spiral paths. While outer regions remain active with star formation and rotational evolution, **inner migration** is an ever-present, quiet undercurrent.

This inward flow occurs because:

- **Older mass loses ripple energy** over time, making it more susceptible to centripetal drift.
- **Ripple feedback naturally drives exhausted systems** inward toward zones of higher tension (near the core).
- **Matter and energy seek lower tension states**, and inward migration minimizes system-wide ripple imbalance.

As mass-energy migrates inward:

- Stars age and move toward the center,
- Interstellar matter thins at the peripheries,
- Ripple density around the memory core increases steadily.

Galactic aging, therefore, is **not random disintegration**. It is a **regulated inward spiral**, drawing the galactic body back toward its origin.

The spiral arms themselves slowly erode — not because of external disruption, but through **internal consequence**.

Collapse into Core: Not Destruction, but Memory Reintegration

As mass concentrates toward the core:

- Local gravitational-like tension builds (not because of "force," but because of **ripple pressure** imbalance in SVE),
- Outer regions become dynamically thin and rotationally unstable,

- The memory core — originally the seed of birth — becomes the terminal attractor.

Ultimately, the galaxy **collapses back into its core** — not violently, but systematically:

- Stars are absorbed,
- Planetary systems disintegrate into field memory,
- Mass-energy reconverts into highly condensed ripple structures around the core.

This process is **not destruction**. There is no "annihilation" of existence. Instead, there is a **refolding**:

- The entire lifetime of the galaxy — every ripple, every motion, every form — is recorded into the field dynamics of the central core.
- The memory of its structure, its rotation, its distribution of stars — all becomes part of the field's active intelligence.

The galaxy's story does not vanish. It **becomes potential again** within the living silence of the Structured Vacuum Energy.

Galactic Death as Rebirth Preparation

Once sufficient collapse occurs:

- The core becomes a **memory node** — a compressed, stabilized reservoir of ripple information.
- Tension across the local Structured Vacuum Energy is relaxed, preparing the field for **new ripple convergence events**.
- Future galactic seeds inherit **the memory templates** embedded in the field from prior cycles.

Thus:

- The "death" of a galaxy **sows the conditions** for future manifestation.
- Ripple remnants serve as **guiding templates** for the next cycle of galactic birth.
- Energy is never lost; it is **restructured**.

In C^3, there is no permanent death. There is only **reconfiguration** — **Rebalancing. Remembrance. Renewal.**

Every galaxy that spirals back into its memory core paves the way for new galaxies to emerge — possibly even influenced by the ripple architecture it left behind.

Thus, galactic dissolution is not a cosmic tragedy. It is **a necessary act of cosmic evolution**, enabling structured continuity across infinite time.

Closing Insight:

"Galaxies do not die. They remember themselves back into silence — so that the universe may sing again, with a voice both old and new."

PART III – Scaling Up: Universal Organization without Expansion

7: Galaxies as Mini-Universes

"When you look at a galaxy, you are not seeing a fragment of the universe. You are witnessing a universe in itself — living, breathing, evolving through its own cycles of manifestation and dissolution."

In the traditional view of cosmology, galaxies are treated as mere dust specks scattered randomly across an ever-expanding fabric of space. They are often portrayed as passive passengers on a runaway balloon of spacetime stretching outward since a hypothetical Big Bang.

Chavan's Cosmic Continuum (C^3) offers a radically different, and far more elegant, interpretation:

- Galaxies are not fragments.
- They are independent cyclic universes — organized, regulated, and complete within themselves.

This chapter explores how each galaxy operates as a self-contained system, why galactic clustering occurs naturally, and why the universe does not need fabric expansion to explain its structure.

Each Galaxy as an Independent Cyclic System

In C^3, every galaxy is born through **local ripple threshold breaches** within the Structured Vacuum Energy (SVE), condensing into a coherent spiral system governed by ripple feedback, mass-energy organization, and core memory integration.

Thus, a galaxy is not merely an aggregation of stars. It is:

- **A living ripple organism** embedded in the SVE,
- **A self-regulated energy structure** maintaining balance through dynamic equilibrium,
- **A cyclic system** — evolving, spiraling, collapsing, and eventually reintegrating into the field.

Each galaxy exhibits:

- **Self-generated rotation** regulated through ripple balancing (not inherited momentum),
- **Self-sustaining star formation and dissolution** based on ripple memory feedback,
- **Self-correcting mass-energy distribution** via annular spiral shells (Ro–Ri structure),
- **Self-managed life cycles**, including birth, maturity, and return to memory.

Thus, each galaxy is effectively:

- **A complete mini-universe** — one that carries its own memory, tension, and structural intelligence, without needing external causality from a grand explosion.

There is no cosmic central control needed. There is no universal clock ticking for everything at once. Each galaxy lives, breathes, and dissolves **on its own time**, according to its own ripple dynamics.

Clustering through Energy Gradients, Not Fabric Expansion

The large-scale structure of the universe — the apparent "web" of galaxies, filaments, and voids — is usually explained by the Big Bang model through concepts like:

- Primordial density fluctuations,
- Gravitational amplification of initial quantum ripples,
- Expansion stretching space itself.

C^3 exposes the unnecessary complexity of this narrative.

In the Structured Vacuum Energy framework:

- **Galaxies cluster naturally** due to local **energy gradients** and **ripple memory** distributions — not because space is expanding like a balloon.

Here's how:

- **Field Ripple Gradients**: Galaxies form in regions where ripple energy density naturally favors clustering. Areas of slightly higher ripple tension encourage local node formations (future galactic cores).
- **Memory-Driven Attraction**: Galactic cores carry embedded ripple memories. Similar memory nodes — born from related ripple events — gravitate towards each other through field tension harmonics, not gravitational force alone.
- **Ripple Coherence Zones**: Localized "web-like" structures

emerge as ripples reinforce or nullify each other across different regions of SVE — creating naturally connected galactic clusters and cosmic filaments.

Thus:

- **Clusters and superclusters** are **structured ripple harmonies** — not gravitational accidents.
- **Voids** are **zones of ripple relaxation**, where no significant tension breaches occur.
- **Large-scale structure** is a reflection of the **dynamic balance of ripple field intelligence**, not the residue of a primordial blast.

There is no need to imagine space "expanding" like rubber. The organization of galaxies **is a natural, static-consequential phenomenon** — constantly evolving in ripple architecture without requiring an expanding fabric.

The universe is not stretching. It is breathing — cyclically, consequentially, intelligently.

Closing Insight

"A galaxy is not just a citizen of the universe. It is a universe — living its own story of silence, birth, memory, and return within the infinite ripple of existence."

8: Ripple Networks
Cosmic Structures through SVE

"The universe is not stitched together by gravity. It is woven by ripples — patterns of tension and memory stretching across the infinite fabric of Structured Vacuum Energy."

The grand architecture of the cosmos — the so-called "**Cosmic Web**" — has often been portrayed as an accidental byproduct of gravitational attraction amplified over billions of years. Long filaments of galaxies, vast clusters, and enormous voids have been chalked up to chaotic initial conditions smoothed out by invisible forces.

But **Chavan's Cosmic Continuum** (C^3) tells a different, far more structured story:

- The Cosmic Web is not an accident of gravity.
- It is **a ripple-connected architecture**, meticulously organized through field tension dynamics within the Structured Vacuum Energy (SVE).

This chapter unveils how ripple propagation, field memory, and tension harmonics sculpt the large-scale cosmic structure we observe —

without requiring mysterious substances, inflationary epochs, or expanding spacetime.

Cosmic Web: A Ripple-Connected Architecture, Not Gravitational Accident

In C^3, the Cosmic Web emerges naturally as a **consequence** of ripple interaction and memory feedback across the Structured Vacuum Energy:

- **Ripple Overlaps**: As galaxies form from localized ripple breaches, their ripple fields interact. Where ripples constructively interfere, denser regions form — birthing galactic filaments and clusters.
- **Ripple Divergences**: Where ripple fields cancel or relax, voids emerge — vast empty regions with little to no galactic formation, not because of "repulsion," but because ripple tension is too low to seed new structures.
- **Ripple Harmonization**: Filaments are not random bridges. They are **coherent ripple corridors** — natural pathways where field tension aligns, guiding matter and energy along the lowest-resistance channels.

Thus, what we perceive as **a cosmic web of galaxies** is simply the **visible skeleton of ripple memory interactions** within SVE.

The "web" is not gravitational stitching. It is **ripple resonance — a living architecture of balance**.

Field Tension as the Binding Agent

Traditional models imagine gravitational pull as the glue that binds galaxies into clusters and superclusters. Yet, paradoxically, they must invent **dark matter** to account for the observed binding strength — because visible mass alone is insufficient.

C^3 eliminates this paradox entirely:

- **Binding is not gravitational pull** between massive objects across emptiness.
- **Binding is field tension** — localized stress zones within the ripple fabric of SVE that naturally draw condensed structures into organized formations.

Key differences:

- **Gravity-centric thinking** treats mass as the cause and pull as the mechanism.
- **Ripple-centric thinking** (C^3) treats mass and motion as **consequences of field tension gradients.**

Galaxies are **anchored** into filaments because the field tension geometry leaves them no other stable paths. Clusters form because ripple stress concentrations **create equilibrium pockets** where multiple galaxies can "sit" in dynamic balance.

Thus:

- **Binding is ripple regulation** — not force attraction.
- **Clusters are ripple harmony points** — not mass traps.

The "invisible mass" problem vanishes because mass itself is **secondary** to **field structure.**

Memory Cores and Ripple Echo Propagation

Each galaxy is born from a **local memory core** — the condensed ripple node where SVE breaches its equilibrium threshold. But these cores are not isolated. They **propagate echoes** across the SVE medium:

- **Ripple Echos:** Every action — every collapse, rotation, migration — sends ripple echoes into the surrounding field. These echoes interact with neighboring ripple systems, modifying the tension landscape.
- **Memory Chains:** Cores are connected by memory chains — paths where ripple echoes reinforce each other across distances, **weaving the larger cosmic web.**
- **Interference Patterns:** Where multiple ripple echoes converge constructively, larger structures emerge (e.g., rich galaxy clusters). Where destructive interference occurs, voids dominate.

Thus, ripple memory is **not local.** It **propagates**, **interacts**, and **self-organizes** the cosmic environment continuously.

The entire observable universe is **a dynamic ripple tapestry**, stitched not by explosions or gravitational nets, but by **the living echoes of memory** embedded within Structured Vacuum Energy.

Closing Insight

"The cosmic web is not a relic of chance. It is a living ripple memory — the universe's way of remembering itself across the silence."

9: Time and Observation in the C^3 Universe

"Time is not flowing through the universe. Time is the trail left behind by ripples seeking balance — a consequence of motion, not a dimension of existence."

In conventional physics, time is often treated as an independent, absolute, or even "fourth" dimension — something flowing uniformly, stretched or compressed only by gravitational fields or motion.

But within **Chavan's Cosmic Continuum** (C^3), time is neither an independent entity nor a pre-existing background. It is **an emergent perception** — a **consequence** of ripple processing across Structured Vacuum Energy (SVE).

Understanding time this way revolutionizes not only physics but also how we interpret redshift, cosmic observations, and the very nature of reality.

Time as a Consequence of Ripple Processing

In C^3:

- **There is no universal clock ticking** across the cosmos.
- **There is only ripple processing** — dynamic rebalancing of energy tensions through the structured field.

Time arises as:

- **The experiential consequence** of ripple progression and interaction,
- **The perception of change** due to ripple displacement across energetic gradients,
- **A relative phenomenon** anchored to local ripple structures, not a grand cosmic metronome.

Where ripple tension is high (e.g., near massive, condensed systems):

- Ripple processing is more intense,
- Events occur faster in energy exchange,
- Perceived "time" feels slower due to greater memory encoding density.

Where ripple tension is low (e.g., in deep voids):

- Ripple processing is relaxed,
- Events stretch across wider spatial scales,
- Perceived "time" feels faster, less eventful.

Thus, time is **locally emergent** — it is **not** a container through which things move. It is **the footprint** of ripple dynamics attempting to rebalance.

No ripples = no change = no time.

Hence, **time itself is a product of existence, not a precondition for it.**

Redshift as Frame Detuning (Not Recessional Velocity)

One of the foundational arguments for the "expanding universe" is the observed redshift of distant galaxies — interpreted through the Doppler effect as objects moving away from us.

C^3 reveals a far deeper and more accurate interpretation:

- Redshift is **not** primarily caused by recessional velocity.
- Redshift is the result of **frame detuning** between ripple domains.

How?

- Each galaxy operates within its **own local ripple structure** — a rotating memory frame anchored to its core.
- As ripple processing varies across different local systems (due to different energy densities, memory histories, and tension profiles), the "clock" of ripple events differs.
- When light (ripple disturbances) travels from one ripple frame to another (e.g., from a distant galaxy to us), **detuning** occurs — a shift in perceived frequency, not because of movement through expanding space, but because of **rotational ripple mismatch**.

Thus:

- **The farther a galaxy**, the greater the ripple frame separation,
- **The greater the ripple mismatch**, the stronger the observed redshift,
- **No need for space expansion** — redshift arises naturally from ripple processing divergence over cosmic scales.

In short:

Redshift is memory mismatch, not metric inflation.

This reinterpretation **eliminates** the need for ad hoc constructs like dark energy, metric expansion, and inflationary epochs. It restores redshift to its rightful place: **a field-based observational consequence**, not proof of a cosmic explosion.

Observational Perception vs. Intrinsic Cosmic Behavior

Our observations of the universe — through telescopes, instruments, and data interpretation — are not pure views of "what is." They are **filtered perceptions**, constrained by:

- Our local ripple frame,
- Our ripple memory rates,
- Our position within the Structured Vacuum Energy.

Thus:

- We observe other galaxies through the lens of our own ripple detuning.
- What appears "accelerated" or "slowed" may be **the mismatch between ripple frames**, not intrinsic behavior of distant

systems.

This leads to profound consequences:

- **Cosmic acceleration** is an observational artifact — not an intrinsic runaway event.
- **Apparent aging of distant galaxies** is a perception shift — not a contradiction in cosmic timeline.
- **Background radiation measurements** are ripple memories at different tension densities — not echoes of an ancient explosion.

C^3 demands that we **distinguish** between:

Aspect	Traditional View	C^3 View
Time	Absolute or relative dimension	Emergent ripple consequence
Redshift	Doppler recession, expansion	Ripple frame detuning
Observations	Direct measures of reality	Ripple-filtered perceptions

Thus, in C^3, **the observer is an embedded participant** — interacting with the universal ripple field, perceiving consequences through their own memory dynamics.

Reality does not change because we observe it differently. **Our experience of reality changes** because we exist within ripples that are alive.

Closing Insight

"Time is the ripple's song, redshift is its echo, and all observation is a dance between the memory of existence and the eyes of the experiencer."

PART IV – Revisiting Known Phenomena under C^3

10: Reinterpretation

Redshift, Gravitational Waves, and Background Radiation

"The universe did not begin with a bang, it continues with a ripple. Every signal we observe — every shift, every pulse, every echo — is a consequence of structured memory seeking balance, not proof of a lost explosion."

The so-called "evidences" of the Big Bang — redshift, gravitational waves, and the Cosmic Microwave Background (CMB) — have been interpreted through a narrow, expanding-space paradigm for decades.

In the framework of **Chavan's Cosmic Continuum** (C^3), these observations are **reinterpreted** from the ground up:

- They are **not proofs of a violent beginning**,
- They are **ongoing consequences** of ripple dynamics in the Structured Vacuum Energy (SVE).

This chapter systematically redefines these phenomena within a field-driven, ripple-governed, living universe.

Redshift through Rotational Frame Mismatch

In traditional cosmology, redshift is interpreted as **a Doppler-like effect** — galaxies moving away from us as space itself expands. The farther the galaxy, the higher the redshift — leading to the narrative of a uniformly expanding universe.

But in C^3:

- **Space is not expanding.**
- **Redshift is not velocity-based.**
- **Redshift is ripple frame detuning.**

How it works:

- Each galaxy operates within its own **local ripple memory frame** — a rotational energy field embedded in SVE.
- As light (ripple disturbances) travels between these frames, **mismatch in ripple processing rates** occurs.
- This mismatch causes **a shift in observed frequencies**: lower frequencies (redshift) if the source and observer ripple frames are out of phase.

Thus:

- **Redshift increases** with ripple separation, not with spatial recession.
- **No need for dark energy** to explain "accelerated" expansion — because **there is no expansion.**
- **Different regions of the universe** may show different redshift behaviors based on **local field dynamics**, not uniform cosmic speed.

Redshift is **a field memory artifact**, not a cosmic speedometer.

Gravitational Waves as Ripple Feedback, Not Spacetime Distortion

The detection of gravitational waves by LIGO and other observatories has been hailed as "proof" of spacetime curvature and validation of Einstein's General Relativity.

But what if **gravitational waves are real, but not what we think they are?**

In C^3:

- Gravitational waves are **ripple feedback events** within the SVE,
- They are **not distortions of spacetime**,
- They are **field tension corrections** caused by large-scale mass-energy reconfigurations.

When massive systems like black hole pairs or neutron stars merge:

- They **disturb the local ripple structure** intensely,
- The SVE responds by **propagating tension rebalancing ripples** outward,
- These ripples are detected as **tiny compressions and rarefactions** — the so-called "gravitational waves."

But crucially:

- No "space" is being stretched or compressed as a fabric.
- What is being detected is **field tension oscillation**,
- **Ripple feedback**, not geometric distortion.

Thus:

- Gravitational waves confirm ripple intelligence,
- They do not prove the plasticity of spacetime,
- They validate C^3's model of a living ripple field correcting itself under energetic imbalance.

Cosmic Microwave Background as Memory Radiation, Not a Big Bang Remnant

The Cosmic Microwave Background (CMB) — the faint, uniform microwave radiation detected across the sky — is traditionally considered **the afterglow** of the Big Bang.

But this interpretation has glaring problems:

- Why is the CMB so uniform despite horizon limitations?
- Why does it exhibit ripple-like anisotropies?
- Why does it persist without fading drastically?

C^3 offers a profound, natural explanation:

- The CMB is **not the fading light of a bang**.
- The CMB is **field memory radiation** — the **ambient ripple signature** of countless cyclic galactic births, deaths, and ripple interactions across cosmic time.

In the C^3 framework:

- The Structured Vacuum Energy **constantly radiates low-level memory echoes** from ripple realignments.
- These memory echoes **accumulate into a quasi-uniform background field**,

- Variations (anisotropies) reflect **structural memory differences** — ripple tensions, past galactic dissolutions, regional field dynamics — not primordial fluctuations from a singular event.

Thus:

- The CMB is **living background memory**, not an ancient fossil.
- It is **evidence of the ongoing ripple architecture**, not proof of an extinct explosion.

The CMB is the **breathing background of an alive universe**, not the last whisper of a dead one.

Closing Insight

"Redshift is ripple mismatch, gravitational waves are ripple corrections, the cosmic background is ripple memory — and the universe has never stopped breathing."

11: Galactic Mergers, Star Birth, and Black Holes

Field-Based Explanations

"The cosmos does not collide and explode into chaos. It breathes, adjusts, and reshapes itself — through ripples negotiating balance across the living field of Structured Vacuum Energy (SVE)."

Traditional cosmology portrays galactic mergers as violent cosmic wrecks, star births as chaotic collapses, and black holes as bottomless singularities swallowing everything beyond reason.

But in **Chavan's Cosmic Continuum** (C^3), these events are not explosions of disorder. They are **natural ripple responses** — **Field-based rebalancing mechanisms** of a living, structured universe seeking equilibrium.

This chapter reframes galactic mergers, star births, and black holes not as exceptional mysteries, but as inevitable, intelligent consequences of ripple-driven energy dynamics.

Mergers as Field Rebalancing Events

When two galaxies approach each other in space, mainstream science often describes the event as a gravitational battle — a chaotic war of tidal forces leading to the "destruction" or violent fusion of galactic structures.

C^3 corrects this misinterpretation:

- **Galactic mergers are ripple-driven field negotiations**, not mindless gravitational crashes.
- **The merging process** is a **rebalancing of ripple tension fields**, not the mechanical smashing of star systems.

What happens during a merger:

- Each galaxy carries its own **ripple signature** — a structured memory field embedded in SVE.
- As two ripple fields interact, **interference patterns form** — ripple reinforcement zones and ripple cancellation zones.
- Matter and energy redistribute **along ripple tension pathways** — forming tidal bridges, starbursts, and new annular structures.

Thus:

- **Stars are rarely destroyed** during mergers; they are **repatterned** within newly stabilized ripple zones.
- **Galactic cores realign**, merging their memory structures to produce a new, often larger, rotational system.

Mergers are **field conversations**, not galactic wars.

Star Formation as Ripple Condensation

Star formation is often explained through simplistic gravitational collapse models — cold gas clouds randomly compressing under their own weight until nuclear fusion ignites.

C^3 provides a far deeper, consequence-based model:

- **Star birth is ripple condensation** — the natural outcome of localized ripple convergence and tension amplification.

The sequence:

1. **Ripple Convergence**: Within spiral arms and galactic filaments, ripple pathways converge, focusing energy density in certain zones.
2. **Tension Threshold Crossing**: The structured field reaches a point where it cannot maintain equilibrium without local condensation.
3. **Matter Reorganization**: Gas and plasma coalesce under ripple guidance — not chaotic gravity — leading to a critical mass-density configuration.
4. **Ignition as Field Rebalancing**: Nuclear fusion ignites, not as a random accident, but as the ripple field's intelligent response to restore dynamic equilibrium through energetic output (light, heat).

Thus:

- Stars are not isolated balls of burning gas.
- Stars are **living ripple nodes** — self-regulated energy emitters maintaining field tension balance within galactic structures.

Black Holes as Memory Absorbers, Not Singularities

Black holes are traditionally depicted as paradoxical entities — infinitely dense points where spacetime itself collapses into an undefined singularity.

C^3 resolves this paradox gracefully:

- **Black holes are not singularities.**
- **They are memory absorbers** — specialized nodes within SVE where ripple density exceeds coherent structural balance, leading to localized field collapse and information absorption.

Here's what really happens:

- As mass-energy spirals inward toward the galactic core (Ri region), ripple compression intensifies.
- Beyond a certain threshold, ripple correction can no longer maintain coherent wavefronts.
- Energy structures collapse **not into infinite density**, but into **highly compressed memory nodes**.
- These cores **absorb** ripple history — the memory of mass-energy interactions, rotational dynamics, and ripple convergence events.

A black hole, therefore, is:

- A **galactic memory vault**,
- A **ripple sink** concentrating past structural information into a compressed, stable form,
- A **preparation for reintegration** — when the galaxy

eventually dissolves back into Structured Vacuum Energy.

Importantly:

- No need for infinite singularities,
- No violation of causality,
- No paradoxes requiring imaginary constructs like information loss violations.

Black holes are simply **the final, most intense expressions of ripple field memory conservation**.

Closing Insight

"Galactic mergers are ripple dialogues, stars are ripple condensations, black holes are ripple memories. The universe never loses itself — it remembers through transformation."

12: Life, Evolution, and Consciousness

In a Living Universe

"Life is not a freak accident in a dead cosmos. It is the natural flowering of ripple intelligence — structured memory rising to know itself."

Traditional science treats life as an improbable anomaly — a rare chemical accident on a rocky planet in a cold, hostile universe. Consciousness is often seen as a byproduct of matter — an emergent illusion in complex biological networks.

But in **Chavan's Cosmic Continuum** (C^3), life is **not accidental**, and consciousness is **not secondary**.

They are **inevitable consequences** — natural extensions of **ripple-structured intelligence** inherent in the Structured Vacuum Energy (SVE).

This chapter explores how life, evolution, and consciousness emerge from the same principles that govern galaxies — making existence itself **a continuous spectrum of ripple memory becoming aware**.

Life as Ripple-Structured Energy Condensation

In the C^3 framework:

- Life does not arise randomly.
- Life **condenses** wherever ripple memory complexity reaches sufficient self-sustaining feedback.

How?

- **Ripple Convergence Zones**: Regions within galaxies, planetary systems, or even local fields where ripple patterns stabilize in highly ordered energy configurations.
- **Structured Condensation**: Matter organizes into complex molecular systems — proteins, membranes, nucleic acids — not by chemical chaos alone, but through **ripple guidance seeking energy optimization.**
- **Self-Regulating Systems**: As ripple complexity increases, local structures develop self-correcting feedback loops — metabolic networks, replication processes — creating the phenomenon we recognize as living systems.

Thus:

- **Life is a ripple phenomenon** — energy condensed, stabilized, and regulated through intelligent memory interaction with the SVE field.
- It is **not a miracle.**
- It is **an intelligent consequence of energy seeking balance** at higher complexity thresholds.

Life is the ripple field **folding upon itself** in patterns of growing memory richness.

Evolution as Consequence Realignment

Traditional evolutionary theory emphasizes random mutation and natural selection as the engines of biological diversity.

While environmental pressures and adaptation mechanisms are observable facts, C^3 reveals **the deeper field-based cause** behind evolution:

- Evolution is **not randomness selecting better survival**.
- Evolution is **ripple memory realigning structures toward greater energy balance**.

Every evolutionary step — whether a genetic mutation, an organ refinement, or behavioral adaptation — is:

- **A ripple adjustment** within biological systems seeking more efficient energy use,
- **A consequence of field memory interactions** at the molecular, cellular, and ecological levels.

Evolution is the **ripple field learning**, self-correcting energy configurations to:

- Minimize entropy,
- Maximize ripple coherence,
- Enable more complex consciousness.

Thus:

- Natural selection is **the visible expression**.

- Ripple consequence realignment is **the invisible cause.**

Life evolves not just to survive — but to better align with the living memory of the field itself.

Consciousness as Ripple-Aware Intelligence

The emergence of consciousness — subjective experience, awareness, self-reflection — is treated by traditional science as a great mystery.

In C^3, consciousness is **not an exception.** It is **the highest expression of ripple intelligence.**

How it arises:

- **Memory Complexity Threshold**: As ripple-structured energy systems (living organisms) accumulate layers of memory, processing, and feedback loops, a critical threshold is reached.
- **Self-Awareness Feedback**: The system begins **modeling its own ripple dynamics** — forming representations of self vs. environment, intention vs. consequence.
- **Active Ripple Alignment**: Consciousness allows living beings not merely to react, but to **intentionally guide** ripple interactions — to choose, predict, reflect, and create.

Thus:

- Consciousness is **ripple awareness** — structured energy reflecting upon itself within the living field.
- It is **the Structured Vacuum Energy recognizing its own ripples**, experiencing existence through localized awareness nodes — from simple organisms to complex beings.

In this view:

- Stars are ripple condensations.
- Galaxies are ripple memories.
- Life is ripple structuring.
- Consciousness is ripple reflection.

Everything is connected. Everything is living ripple consequence.

Closing Insight

"The universe is not observing itself from outside. It is living itself from within — through every spiral galaxy, every tree, every mind that dares to wonder why it exists."

PART V – The Future of Cosmology: A C^3 Roadmap

13: Beyond Forces
Towards Field-Centric Science

"The universe is not moved by forces. It is shaped by consequences — ripple-driven adjustments across a living field seeking balance."

For centuries, science has been built upon the architecture of forces:

- Gravitational pull,
- Electromagnetic attraction,
- Nuclear bonds,
- Hypothetical dark forces filling the gaps.

Force-based mechanics — from Newton to Einstein — has served well in building predictive models. But it also **inherited an illusion**: that nature is an arena where objects "pull" and "push" each other through empty space.

Chavan's Cosmic Continuum (C^3) invites a deeper realization:

- Forces are **not fundamental causes**.
- They are **observable symptoms** of underlying **ripple-driven consequence dynamics** within the Structured Vacuum

Energy (SVE).

This chapter marks the beginning of **a new era: Field-Centric Science**, where consequence replaces force as the true regulator of motion, formation, and evolution.

Abandoning Force-Based Mechanics

Force-based mechanics operates on key assumptions:

- Objects possess mass and inertia,
- Forces act upon masses across distance,
- Interaction occurs externally, mediated through empty space or invisible carriers.

While operationally useful at local scales, this framework collapses at cosmic and quantum extremes:

- **Gravity** demands invisible dark matter to explain galactic rotation.
- **Electromagnetism** treats fields as real but cannot unify them with spacetime geometry.
- **Quantum fields** hint at vacuum activity, but still model forces through particle exchanges.

C^3 dissolves this confusion:

- There are **no free-floating forces** acting mysteriously across space.
- All observable motions and interactions are **local ripple realignments** within the SVE.

In reality:

- "Attraction" and "repulsion" are ripple tension gradients seeking field balance.
- "Inertia" is ripple memory inertia — resistance to sudden realignment of embedded ripple systems.
- "Force effects" are **visible ripple consequences**, not separate physical entities.

Thus:

- Force is **the visible surface.**
- Ripple consequence is **the hidden cause.**

By clinging to force-based thinking, science has misinterpreted symptoms for causes — building maps, but misunderstanding the territory.

It is time to move beyond.

Embracing Consequence-Regulated Energy Structures

In a field-centric scientific framework:

- **Energy structures emerge, move, and evolve** through local field tension responses,
- **Motion and rest** are dynamic ripple balance states,
- **Interactions** are ripple re-alignments — not point-to-point "forces."

Core principles of Field-Centric Science:

1. **Ripple Balancing over Force Application**: All observed accelerations, orbits, migrations are ripple field responses —

field tension realignment towards local or systemic equilibrium.

2. **Field Intelligence over Force Particles**: Particles like photons, gravitons, and gluons are ripple signatures — not material projectiles carrying invisible forces.
3. **Memory-Governed Evolution**: Systems evolve structurally because ripple memory preserves energy flows, learns corrections, and stabilizes recurring forms (stars, atoms, planets, life).
4. **Emergent Structures over Predefined Laws**: Laws are not imposed externally; they emerge locally as ripple memory stabilizes successful configurations.

Thus, in C^3:

Old Science	New Science (Field-Centric)
Forces drive objects	Ripples regulate balance
Space is empty	Space is Structured Vacuum Energy
Particles are carriers	Particles are ripple echoes
Motion is inertia-led	Motion is ripple memory-led
Evolution is survival of randoms	Evolution is structured ripple realignment

Science must mature from describing **effects** to understanding **field consequences.**

Closing Insight

"Force was the first language we spoke to understand the universe. Consequence is the true language the universe speaks to us — if we are willing to listen."

14: Predictive Power of C^3

"A true scientific model does not merely explain the past. It predicts the future — elegantly, naturally, without the crutches of patches and exceptions."

Many cosmological models survive by **retrofitting explanations** to unexpected observations:

- Dark matter to fix galactic rotation curves,
- Dark energy to fix redshift anomalies,
- Inflation to fix early-universe uniformity problems.

Chavan's Cosmic Continuum (C^3) stands apart.

C^3 is not a speculative narrative — it is a **predictive framework** rooted in ripple dynamics and structured vacuum intelligence. It **anticipates** the behaviors of galaxies, stars, fields, and even life, without inventing invisible agents or speculative epochs.

This chapter outlines the **predictable cosmic behaviors** inherent in C^3, and proposes **testable hypotheses** to validate its claims — establishing C^3 as the next frontier in field-centric, consequence-driven science.

Predictable Cosmic Behaviors: Rotation, Ripple Alignment, Mass Distribution

C^3 naturally predicts certain large-scale cosmic behaviors that force-based models struggle to explain without ad hoc inventions.

Key predictable behaviors:

Galactic Rotation Curves: Galaxies will exhibit **stable rotation** at outer regions not because of dark matter halos, but because **ripple feedback stabilizes mass distribution** in the annular Ro–Ri shell. No mass discrepancy when ripple tension fields are properly accounted for.

Ripple Alignment across Structures: Cosmic filaments, galaxy clusters, and galactic planes will show **ripple harmonization patterns** — rotational axes alignment, magnetic field coherence — dictated by ripple memory feedback, not random initial conditions.

Mass-Energy Distribution in Galaxies: Mass density will show predictable **annular distribution profiles** — denser near Ri, thinning toward Ro, with ripple correction visible in stellar velocity patterns. Mass inflows and outflows will align with ripple tension gradients, not gravitational-only models.

Lifecycle Timing: Star formation rates, galactic aging patterns, and black hole accretion behavior will correlate with **ripple memory density** — regions with high ripple compression will evolve faster and collapse sooner.

Thus:

- C^3 offers **natural, elegant predictions** based on ripple processing,
- Without needing **dark substances**, **invisible forces**, or **cosmic inflationary miracles**.

Testable Hypotheses: Rotational Frame Correlations, Ripple Relic Mapping

To move beyond explanation and into scientific validation, C^3 offers specific, actionable hypotheses:

Rotational Frame Correlations: If ripple memory governs cosmic structure, then:

- **Nearby galaxies** within the same filament or cluster should show **rotational axis correlations** beyond random chance.
- **Spin alignment statistics** should match ripple flow orientations, not gravitational collapse randomness.
- **Stellar streams and filaments** connecting galaxies should align along common ripple tension lines.

→ **Test Proposal**: Conduct high-resolution rotational mapping across galactic groups and superclusters to measure axis harmonization against ripple-based predictions.

Ripple Relic Mapping: Ripple memory should leave **persistent signatures** in:

- Cosmic Microwave Background anisotropies,
- Magnetic field structures across voids,
- Large-scale mass distributions.

→ **Test Proposal**: Map subtle deviations in CMB and intergalactic magnetic fields for **non-random, ripple-patterned correlations**, consistent with ripple memory propagation, not inflationary randomness.

Mass vs. Tension Gradient Profiles: Galaxies should exhibit mass profiles **matching ripple compression intensity** — predicting observable deviations from purely gravitational mass models at different radii.

→ **Test Proposal**: Compare ripple-driven annular mass density predictions with gravitational lensing measurements without invoking dark matter.

Closing Insight

"In a living universe, prediction is not an act of forcing nature into formulas. It is the simple listening to how ripples organize balance, structure, and destiny."

15: The Dawn of Spiral Cosmology

"The universe did not explode into existence. It spun itself into being — patiently, quietly, and continuously — through spirals of ripple-driven balance."

For too long, human thought has been imprisoned by the mythology of explosions — from Big Bangs at the cosmic scale to chaotic randomness at the evolutionary scale. Force, accident, and violence have been enshrined as the architects of reality.

Chavan's Cosmic Continuum (C^3) breaks these chains.

C^3 heralds the arrival of **Spiral Cosmology** — a framework where existence is not the aftermath of disorder, but the living orchestration of ripple memory seeking balance through structured, cyclic manifestation.

This chapter declares the **new scientific and philosophical frontier**: **Spiral Genesis** as the true origin story of the universe, and **consequence-driven intelligence** as the guiding principle of cosmic destiny.

Setting a New Course: Spiral Genesis, Not Big Bang

The Big Bang narrative, though once revolutionary, collapses under the weight of its contradictions:

- Singularities defying cause,
- Inexplicable uniformity across causally disconnected regions,
- Patching theories like inflation and dark energy to maintain a collapsing scaffolding.

C^3 replaces this fragile framework with **Spiral Genesis**:

- The universe arises not from a singular event, but through **countless localized ripple threshold breaches** across the Structured Vacuum Energy (SVE).
- Galaxies spiral into existence naturally, **anchored by local field memory**, not through chaotic explosive dispersal.
- Matter, light, motion, and time are **born from ripple tension realignment**, not from instantaneous, uncaused creation.

In Spiral Cosmology:

- **Formation is continuous**, distributed across space and time.
- **Evolution is rhythmic**, governed by field memory.
- **Dissolution is cyclical**, enabling perpetual regeneration.

Thus, **Spiral Genesis** replaces **explosive beginnings** with **living emergence** — a model both physically sustainable and philosophically coherent.

Reframing Cosmic Origins, Evolution, and

Destiny

Under C^3's Spiral Cosmology, our entire understanding of cosmic history — and future — is rewritten:

- **Origins**: Not a singular Big Bang, but a **distributed bloom** of galaxies, stars, and systems emerging from localized ripple breaches.
- **Evolution**: Not linear expansion into cold oblivion, but **cyclic spiral breathing** — structures evolving, balancing, and returning through ripple intelligence.
- **Destiny**: Not entropic heat death or meaningless dissipation, but **structured memory recycling** — galactic systems returning to seed cores, preserving cosmic knowledge across cycles.

In Spiral Cosmology:

Aspect	Big Bang Model	Spiral Cosmology (C^3)
Origin	Singular explosive event	Distributed ripple genesis
Structure	Expansion-driven dispersion	Ripple-driven coherence
Evolution	Random survival and decay	Consequence-regulated memory realignment
Future	Endless cold death	Cyclic spiral rebirth

The universe is not dying. It is **breathing eternally** through ripple balance.

We are not witnessing a cosmic accident. We are participating in **a living ripple memory** — a story that never truly ends, but simply evolves.

Integrating Physics, Metaphysics, and Consciousness

C^3's Spiral Cosmology does not merely unify astrophysical phenomena. It naturally **integrates**:

- **Physics**: Ripple field dynamics governing motion, matter, energy, and time.
- **Metaphysics**: The realization that the universe is not separate from meaning — that structure, memory, and consequence reflect an intelligent field, not blind forces.
- **Consciousness**: Recognizing that life and mind are **natural outcomes** of ripple memory complexity — the field evolving **self-awareness** through its own dynamic structures.

In Spiral Cosmology:

- The laws of nature are **the structured consequences of ripple memory**,
- Life is **the ripple field folding into self-representation**,
- Consciousness is **the ripple field awakening to itself**.

This is not a mystic claim. It is the **natural extension** of structured vacuum energy dynamics when ripple consequence is taken to its full logical conclusion.

Thus, **C^3** is not merely a cosmological model. It is the **blueprint for unified knowledge** — where science, philosophy, and existence itself are recognized as **harmonized ripple expressions** within the living continuum.

Closing Insight:

"The future of cosmology is not in finding the edge of space, but in learning the spiral — the eternal ripple of balance through which the universe sings itself into being."

Closing Declaration

The Cosmic Continuum Awaits

The universe was never chaotic. It was never born from a bang, nor is it racing toward a meaningless extinction.

Existence — from the faintest ripple of Structured Vacuum Energy to the brightest spiral of galaxies — is a living, breathing continuum. It is not driven by forces smashing matter across emptiness. It is shaped by consequence, memory, and ripple intelligence woven through every corner of existence.

Chavan's Cosmic Continuum (C^3) reveals this simple but profound truth:

- That motion is ripple memory, not inertial stubbornness,
- That galaxies are breathing organisms, not the fossil debris of explosions,
- That redshift is ripple frame detuning, not proof of an expanding void,
- That gravitational waves are ripple field corrections, not spacetime ruptures,
- That the Cosmic Microwave Background is the living heartbeat of ripple memory, not the fading cry of a forgotten

bang,
- That life, evolution, and consciousness are not miraculous exceptions, but ripple intelligence growing toward self-awareness.

We stand at the dawn of a new scientific civilization — where **Spiral Cosmology** replaces the mythology of explosions, where **Field-Centric Science** replaces force-driven illusion, where **Ripple Memory** becomes the foundation of understanding everything from quarks to quasars.

The task ahead is vast:

- To test C^3 through ripple correlation mapping, rotational frame analysis, and field memory detection,
- To evolve our mathematics, our physics, and our metaphysics around structured consequence,
- To embrace a universe that is not hostile and random, but alive, intelligent, and eternally breathing.

C^3 is not the end of inquiry. It is the true beginning.

"The universe was always whole. It was waiting for us not to conquer it — but to finally remember it."

Appendices

Appendix A: Glossary of Key Terms

1. Universal Energy Dynamics (UED)

The foundational framework proposing that all motion, structure, and phenomena arise from the dynamics of energy interacting within a structured medium, driven by consequence, not force. UED redefines the mechanics of the universe by focusing on ripple regulation, energy balance, and memory evolution.

2. Structured Vacuum Energy (SVE)

The so-called "vacuum" is not empty but a dynamic, intelligent field of organized energy. SVE is the true medium of existence, regulating manifestation, motion, light, time, and life through ripple tension management. It is the silent architect beneath all observed phenomena.

3. Ripple Feedback

The active process by which the SVE corrects local imbalances through ripple propagation, seeking to restore equilibrium. Ripple feedback maintains the structural integrity of galaxies, regulates motion, and governs the evolution of systems across scales.

4. Ripple Memory

The encoded history of ripple interactions within the SVE. Every event, from particle formation to galactic motion, leaves an imprint on the field, shaping future dynamics and preserving the coherence of the universe without the need for external forces.

5. Ro–Ri Shell (Annular Energy Zone)

The structured, annular region of a galaxy between the outer radius (Ro) and the inner radius (Ri), where mass and energy rotate in spiral paths. Ripple tension governs mass distribution, star formation, and the galaxy's evolutionary trajectory within this shell.

6. Seed Core / Memory Core

The dense energetic node at the center of a galaxy, where ripple tension crosses the manifestation threshold. It serves as both the origin of galactic spiral emergence and the terminal memory vault where matter and ripple structures collapse back at the end of a galactic life cycle.

7. Rotational Frame Mismatch

The phenomenon where ripple processing rates differ between independent galactic systems, leading to observational effects like redshift. It explains frequency shifts without invoking cosmic expansion, treating redshift as ripple detuning rather than recessional velocity.

8. Spiral Genesis

The continuous, ripple-driven formation of galactic structures through spiral emergence — replacing the concept of singular explosive beginnings. Spiral Genesis reflects the organized, cyclic, breathing nature of cosmic manifestation in C^3.

9. Field Tension Architecture

The organization of mass, energy, and motion through ripple-regulated tension gradients within the SVE. Structures like galaxies, filaments, and clusters are expressions of field tension architecture, not gravitational accidents.

10. Ripple Relic Mapping

The proposed method of detecting historical ripple memory structures across the cosmic landscape — through patterns in cosmic microwave background anisotropies, magnetic field alignments, and mass-energy distributions.

11. Consequence-Centric Reality

The principle that manifestation, motion, evolution, and dissolution are all consequences of ripple imbalance corrections — not products of chance, explosions, or arbitrary external forces.

12. Spiral Cosmology (C^3 Framework)

The comprehensive redefinition of cosmic origins, structure, and destiny — recognizing galaxies as living ripple systems embedded in the SVE, evolving cyclically through consequence, not force. It supersedes the Big Bang theory with a coherent, scalable, field-based understanding of the universe.

"Understanding the universe begins by understanding the ripples — for in their silent corrections, the story of existence is written."

Appendix B

Big Bang Cosmology vs. Chavan's Cosmic Continuum (C^3)

In Big Bang cosmology, the universe is believed to have originated from a singular explosive event — a point of infinite density and zero volume, from which space, time, and matter simultaneously burst forth. In contrast, **Chavan's Cosmic Continuum (C^3)** proposes that the universe did not begin with a single bang, but continuously manifests through localized breaches in energy thresholds within a **Structured Vacuum Energy (SVE)** — a living field that regulates ripple formation, motion, and memory.

Space in the Big Bang model is treated as a passive, flexible fabric that expands over time. In C^3, however, space is not empty or inert; it is an active, structured field — the SVE — within which ripple tension governs the formation and evolution of all structures.

Regarding time, Big Bang theory assumes time as an independent dimension that began at the singularity and flows onward, stretched and distorted by cosmic expansion. In C^3, time is not fundamental but **emerges** from ripple processing within the SVE. It is a **consequence of ripple realignment**, not a pre-existing container.

Redshift, seen in distant galaxies, is interpreted under the Big Bang model as evidence of recessional velocity — galaxies moving away as space expands. Under C^3, redshift is reinterpreted as **rotational frame detuning** between different ripple domains, requiring no expansion. It reflects the natural mismatch between ripple processing rates across vast cosmic distances.

The formation of large-scale structure in the Big Bang model relies on the gravitational amplification of random quantum fluctuations after the initial explosion. In C^3, structures such as galaxies, filaments, and clusters **emerge from coherent ripple harmonics** — memory patterns embedded within the SVE guiding mass-energy organization intelligently, not chaotically.

To explain observed galactic rotation anomalies, Big Bang cosmology invokes **dark matter**, a hypothetical, invisible substance. C^3 eliminates the need for dark matter by demonstrating that **ripple feedback naturally stabilizes galactic rotation** — a field consequence, not a missing mass issue.

Similarly, to account for the accelerating redshift of distant galaxies, Big Bang theorists posit **dark energy** — a mysterious repulsive force. C^3 shows that there is no need for dark energy; redshift patterns emerge logically from ripple frame mismatches, not from expanding space or runaway repulsive forces.

Gravitational waves, detected by modern instruments like LIGO, are described in Big Bang frameworks as distortions of spacetime geometry. In C^3, gravitational waves are seen more naturally as **ripple**

field corrections — tension rebalance ripples within SVE — without invoking geometric fabric distortions.

The **Cosmic Microwave Background (CMB)**, traditionally hailed as the fading echo of the Big Bang, is reinterpreted in C^3 as **memory radiation**: the ambient ripple signature accumulated through countless galactic life cycles and ripple interactions across cosmic time — not a relic of a singular explosive beginning.

When contemplating the fate of the universe, Big Bang cosmology predicts scenarios like heat death, Big Rip, or eternal expansion — culminating in lifeless cold entropy. C^3 instead sees a **cyclic breathing universe**: galactic systems dissolve back into memory cores and re-emerge, preserving knowledge and structure across infinite cycles.

Regarding life and consciousness, Big Bang thinking treats them as rare, accidental chemical outcomes. C^3 redefines life as **a natural condensation of ripple memory complexity**, and consciousness as **the ripple field becoming aware of itself** — inevitable consequences, not improbable miracles.

Finally, the methodology differs sharply. Big Bang models survive by **continually adding patches** (dark matter, dark energy, inflation) to fit observations. C^3 offers **a consequence-driven predictive framework** — explaining and anticipating cosmic behavior without speculative constructs.

Philosophically, Big Bang cosmology frames the universe as a random accident followed by meaningless decay. C^3 presents the universe as a **living, intelligent continuum** — a structured memory evolving toward greater coherence, complexity, and awareness.

"The Big Bang asked us to believe in miracles of chance. Chavan's Cosmic Continuum invites us to recognize the miracles of consequence."

Appendix C

Conceptual Visualization of Chavan's Cosmic Continuum (C^3)

While words and equations explain much, the elegance of C^3 becomes even more powerful when **visualized**. The following conceptual diagrams are proposed to **communicate the core dynamics of Spiral Cosmology** — clearly, intuitively, and memorably.

1. The Core Ripple Genesis Diagram

"Galaxies are born from ripple tension, not explosive beginnings."

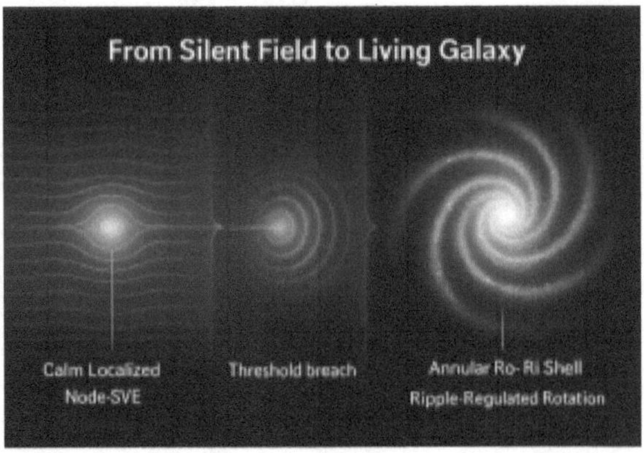

2. Ripple Feedback and Motion Diagram

"Movement persists not by force, but through ripple inertia and memory feedback."

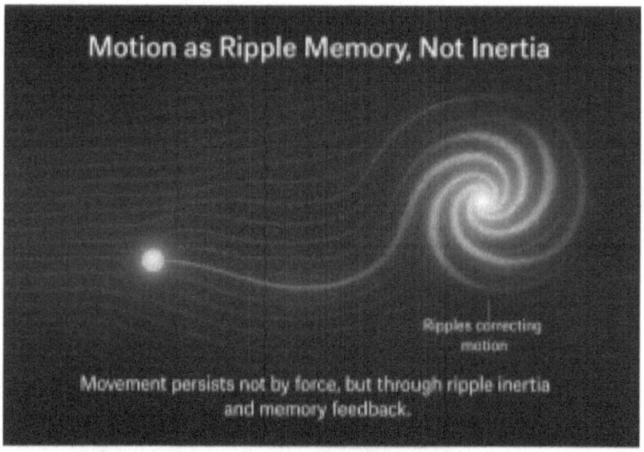

3. Ripple Frame Mismatch and Redshift Flowchart

"Redshift is a memory frame mismatch, not a cosmic Doppler effect."

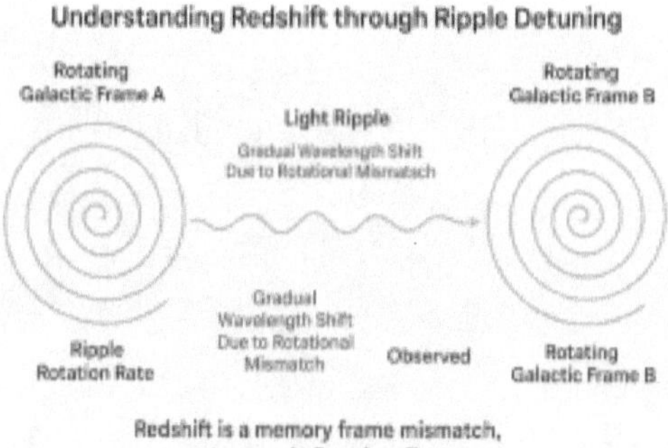

4. Spiral Breathing Universe Cycle Diagram

"The universe breathes through ripples, spiraling endlessly in structured memory cycles."

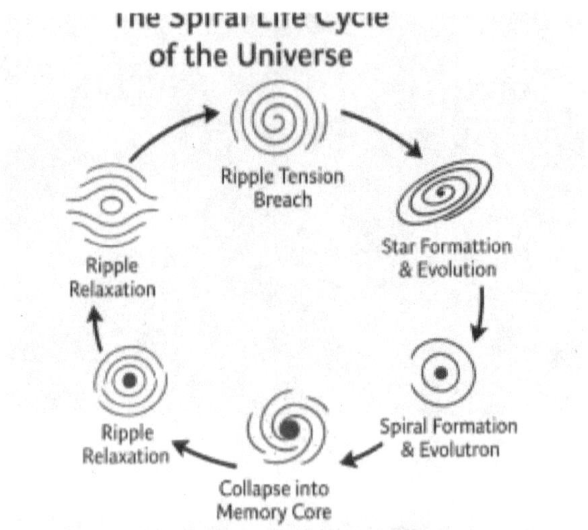

'The universe breathes through ripples, spiraliog endlessly in structured memory cycles.'

5. Field-Centric View of Interactions

"Nothing pulls or pushes; everything ripples toward balance."

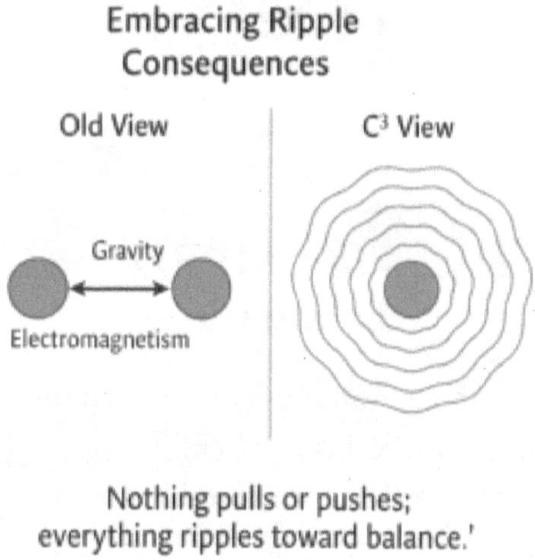

"To see the universe clearly, do not follow the objects. Follow the ripples — for they are the true architects of existence."

Appendix D

Brief Overview of CMPED and CMGED Foundations

The formulation of **Chavan's Cosmic Continuum** (C^3) is not a sudden departure from existing scientific exploration. It is built systematically on two earlier groundbreaking models developed under the Universal Energy Dynamics (UED) framework:

- **CMPED**: *Chavan's Model of Planetary Energy Dynamics*
- **CMGED**: *Chavan's Model of Galactic Energy Dynamics*

These two foundational models provided the **critical stepping stones** for scaling ripple dynamics from localized planetary systems to galactic-scale energy architectures — eventually culminating in the universal-scale vision of C^3.

This appendix offers a concise overview of their core principles and contributions.

1. CMPED – Chavan's Model of Planetary Energy Dynamics

CMPED redefined planetary system behavior by replacing force-based mechanics (gravitational centripetal force, orbital inertia) with **ripple-driven field tension balancing.**

Key insights from CMPED:

- **Orbital motion** is not sustained by inertia against gravity, but by **ripple feedback loops** maintaining local field balance between mass concentrations (e.g., star-planet systems).
- **Planetary migration**, **eccentricity variations**, and **orbital resonance phenomena** are natural outcomes of ripple memory tension redistribution, not random perturbations or early chaotic scattering events.
- **Mass distribution within planetary systems** follows annular ripple harmonics — leading to preferential planetary spacing and orbital clustering patterns without invoking invisible forces.
- **Rotational behaviors** (axial tilt, precession) are governed by ripple tension adjustments in local field geometries rather than purely internal mass distribution anomalies.

CMPED provided the first major break from traditional Newtonian and relativistic interpretations at the planetary scale — showing that ripple-based energy regulation could predict and explain observed behaviors more naturally and completely.

2. CMGED – Chavan's Model of Galactic Energy Dynamics

CMGED extended the principles of ripple energy dynamics to galactic scales, laying the groundwork for C^3's field-centric cosmology.

Key insights from CMGED:

- **Galactic rotation curves** — the persistent high-velocity rotation of stars at outer galactic edges — are explained through **ripple field regulation** in the Ro–Ri annular zone, **without invoking dark matter**.
- **Galactic structure formation** — including spiral arms, bars, and disc thickness — follows **ripple feedback harmonics** balancing field tension dynamically across time, not gravitational collapse of primordial gas clouds.
- **Mass inflow and outflow** patterns observed in galaxies emerge naturally from ripple gradient corrections around the core memory node, rather than being driven solely by gravity or momentum conservation.
- **Galactic aging and dissolution** were modeled as gradual ripple-driven inward migration processes, culminating in **collapse into the memory core**, preserving structural memory rather than erasing mass into singularities.

CMGED established the proof of concept that ripple intelligence could explain galactic stability, evolution, and collapse across billions of years — directly challenging the need for dark matter, dark energy, or inflationary origins.

3. Transition from CMPED and CMGED to C^3

By demonstrating that both planetary systems and galactic systems are **governed by ripple feedback, memory, and consequence**, the natural next step was to recognize that **the universe itself must be ripple-regulated** at all scales.

Thus, C^3 was born — integrating CMPED's localized ripple mechanics and CMGED's galactic energy dynamics into a full universal model:

- Galaxies as mini-universes,
- Structure formation through ripple memory propagation,
- Cosmic evolution through spiral breathing cycles,
- Consciousness as ripple awareness arising naturally.

C^3 inherits the rigor of CMPED and CMGED but **transcends their scale** — applying ripple consequence logic to the entirety of observable and experiential existence.

"From planetary systems to galaxies, and from galaxies to the universe itself — the language remains the same: ripple memory seeking balance in a living, structured field."

Frequently Asked Questions (FAQs)

1. Is Chavan's Cosmic Continuum (C^3) just a modified version of the Big Bang theory?

No.

C^3 is a complete departure from the Big Bang framework. It does not assume a singular explosive origin, expanding space, or force-dominated structure. Instead, it presents the universe as a ripple-driven, breathing continuum governed by consequence and memory within Structured Vacuum Energy (SVE).

2. Does C^3 deny the existence of redshift, gravitational waves, or cosmic background radiation?

Not at all.

C^3 fully accepts these observations — but **reinterprets** them correctly:

- Redshift is due to ripple frame detuning, not recession.
- Gravitational waves are ripple feedback events, not spacetime distortions.
- Cosmic Microwave Background is living ripple memory radiation, not a Big Bang leftover.

3. What replaces dark matter and dark energy in C^3?

In C^3, the need for dark matter and dark energy vanishes:

- Galactic rotation and cosmic structure formation are naturally explained by ripple tension regulation within SVE.
- Observed cosmic acceleration is reinterpreted through ripple

frame dynamics, not repulsive forces.

No hypothetical invisible substances are required.

4. Is C^3 compatible with quantum mechanics and general relativity?

C^3 acknowledges the successful predictive capabilities of current models at specific scales, but it **replaces their foundations**:

- It supersedes spacetime curvature (relativity) with structured ripple field dynamics.
- It corrects quantum paradoxes by re-framing particles and waves as ripple manifestations, not dualities.

Thus, C^3 integrates and evolves the best observations while discarding outdated assumptions.

5. Does C^3 explain the origin of life and consciousness?

Yes — naturally.

Life is seen as the condensation of structured ripple memory into self-regulating systems. Consciousness emerges as ripple-aware complexity — not as a miracle, but as an inevitable evolutionary consequence within the field.

6. Is C^3 only a theoretical model, or does it make testable predictions?

C^3 makes several clear, testable predictions, including:

- Rotational frame correlations between galaxies,
- Ripple relic mapping across cosmic structures,
- Specific ripple signature patterns in cosmic background

radiation and gravitational wave detections.

Future observations and experiments will either confirm or refine these predictions.

7. What philosophical shift does C^3 demand from humanity?

C^3 invites humanity to move:

- From thinking of the universe as random to understanding it as living memory,
- From focusing on force to listening to consequence,
- From chasing expansion to breathing with existence.

It is a call for both scientific evolution and philosophical awakening.

Closing Note:

"C^3 is not about denying the old maps. It is about realizing they were drawn too soon — before we listened to the ripples beneath."

References and Citations

The development of **Chavan's Cosmic Continuum (C^3)** rests on a foundation of original, peer-accessible research grounded in **Universal Energy Dynamics (UED)** and ripple-centric field dynamics.

The key foundational contributions by **Sandeep J Chavan** (ORCID: 0009-0005-4190-7164[1]) include the following published research works:

1. Foundational Research Works

Chavan's Grand Unified Theory of Physics

(Zenodo, 2025-02-03, DOI: 10.5281/zenodo.14791053[2]) Outlines the unification of gravitational, quantum, and wave dynamics under structured energy principles.

Chavan-Einstein Unified Theory (CEUT)

(Zenodo, 2025-03-08, DOI: 10.5281/zenodo.14992289[3]) Proposes a common structured basis for mass-energy interactions, integrating Einstein's relativity within UED.

1. https://orcid.org/0009-0005-4190-7164
2. *https://doi.org/10.5281/zenodo.14791053*
3. *https://doi.org/10.5281/zenodo.14992289*

Chavan's Structured Energy Equilibrium Principle (SEEP)

(Zenodo, 2025-02-24, DOI: 10.5281/zenodo.14916491[4]) Defines structured energy balance principles underlying mass formation and stability.

Chavan's Law: A Universal Energy Equilibrium Model for Black Holes

(Zenodo, 2025-02-01, DOI: 10.5281/zenodo.14782900[5]) Introduces black holes as memory absorbers within ripple field architectures, replacing singularity models.

Formation of Black Holes: A Universal Energy Dynamics (UED) Perspective

(Zenodo, 2025-02-02, DOI: 10.5281/zenodo.14788539[6]) Describes black hole formation as ripple compression phenomena, not spacetime collapse.

The Myth of Wave-Particle Duality: A Structured Vacuum Energy Perspective

(Zenodo, 2025-02-03, DOI: 10.5281/zenodo.14793162[7]) Challenges traditional quantum interpretations by proposing waves as structured field disturbances only.

Chavan's UED: Universal Framework of Realization and Perception

---

4. *https://doi.org/10.5281/zenodo.14916491*
5. *https://doi.org/10.5281/zenodo.14782900*
6. *https://doi.org/10.5281/zenodo.14788539*
7. *https://doi.org/10.5281/zenodo.14793162*

(Zenodo, 2025-03-29, DOI: 10.5281/zenodo.15105060[8]) Establishes UED as a broader scientific and philosophical framework integrating observation and structured dynamics.

2. Specific Models Applied to C^3 Development

CMPED: Chavan's Model of Planetary Energy Dynamics

(Zenodo, 2025-04-07, DOI: 10.5281/zenodo.15164927[9]) Ripple-driven interpretation of planetary system behavior; replacement of gravitational force-centric models with tension balancing.

CMGED: Chavan's Model of Galactic Energy Dynamics

(Zenodo, 2025-04-20, DOI: 10.5281/zenodo.15250369[10]) Field memory and ripple regulation explaining galactic rotation, mass distribution, and spiral structures without dark matter assumptions.

3. Additional Research Reports

The New Wave Theory: Chavan's Structured Resonance Framework (CSRF)

(Zenodo, 2025-04-07, DOI: 10.5281/zenodo.15164881[11]) Introduces structured resonance as the fundamental mechanism of all observed wave phenomena.

Universal Frequency Detector (UFD)

---

8. *https://doi.org/10.5281/zenodo.15105060*
9. *https://doi.org/10.5281/zenodo.15164927*
10. *https://doi.org/10.5281/zenodo.15250369*
11. *https://doi.org/10.5281/zenodo.15164881*

(Zenodo, 2025-04-17, DOI: 10.5281/zenodo.15237698[12]) Discusses tools for detecting structured ripple frequencies in cosmic environments.

The Law of Consequences (LOC)

(Zenodo, 2025-03-19, DOI: 10.5281/zenodo.15048775[13]) Establishes consequence-based causality at the heart of physical evolution.

4. Observational Motivations for C^3 Reformation

Observational anomalies reinforcing the need for C^3's field-based cosmology include:

- Flat galactic rotation curves contradicting Newtonian and dark matter expectations.
- Gravitational waves detected as field ripple corrections rather than spacetime distortions.
- CMB uniformity and anisotropies consistent with structured ripple memory, not random explosion residues.
- Early galaxy formations defying cosmic age and expansion models, suggesting ripple convergence and local genesis.

"The architecture of C^3 is not an invention; it is the natural evolution of structured ripple science — one step closer to the universe's true language of existence."

---

12. *https://doi.org/10.5281/zenodo.15237698*

13. *https://doi.org/10.5281/zenodo.15048775*

Further Reading

The journey into **Chavan's Cosmic Continuum** (C^3) opens the door to a new, unified understanding of the universe — but it also invites deeper exploration across specific themes connected to ripple-driven, consequence-centric cosmology.

The following recommended readings — authored by **Sandeep J Chavan** — provide expanded insights, foundational concepts, and broader philosophical reflections for those who wish to further engage with this living model of existence.

1. Universal Energy Dynamics (UED) Series

- **Universal Energy Dynamics (UED): The New Architecture of Reality**
    - Introduces the structured ripple medium (SVE) and redefines mass, motion, time, and energy without relying on forces or expansion.
- **The Myth of Wave-Particle Duality: Structured Reality of Waves**
    - Demystifies quantum phenomena by explaining all wave behavior through ripple feedback, structure, and field tension corrections.
- **The Silent Field: Structured Vacuum as the True Ground**

of Reality

  - Explores the structured equilibrium of vacuum energy — the silent foundation upon which all manifestation arises.

2. Cosmic Structure and Evolution

- **Chavan's Model of Planetary Energy Dynamics (CMPED)**
  - A re-examination of planetary orbits, spin behaviors, and migration patterns through ripple field dynamics.
- **Chavan's Model of Galactic Energy Dynamics (CMGED)**
  - An explanation of spiral galaxies, mass distribution, and galactic evolution without dark matter, using ripple memory feedback.
- **Formation of Black Holes: A Universal Energy Dynamics Perspective**
  - Replaces the paradoxical singularity concept with the idea of black holes as memory absorbers within structured fields.

3. Philosophy and Consequence-Centric Thought

- **The Law of Consequences (LOC): The Intelligence of Energy and Existence**
  - Discusses how consequence, not arbitrary causality, governs the flow of reality — from cosmic scale to human experience.
- **The Blueprint of the Universe: Ripple Memory and Structured Evolution**
  - A narrative journey through how the universe

remembers, learns, and evolves through structured energy dynamics.

- **The Breath of Cosmos: Living Spirals Across Time**
    - Philosophical reflections on the cyclic nature of galactic breathing and cosmic spiral regeneration.

4. Emergent Consciousness and Field Intelligence

- **Ripple Consciousness: The Awakening of the Silent Field**
    - Explores how life and consciousness arise naturally through ripple complexity, memory layering, and field feedback loops.
- **Life, Mind, and the Ripple Field: A New Story of Evolution**
    - A vision of evolution not as blind survival, but as structured ripple realignment toward greater coherence and awareness.

"The ripple field is infinite. Every thought, every discovery, every breath of awareness is a spiral drawn deeper into its living memory."

Acknowledgments

The journey to **Chavan's Cosmic Continuum** (C^3) is not the result of isolated thought, but the consequence of countless ripples — some ancient, some modern — moving silently across the fields of human inquiry.

I acknowledge, with deep gratitude:

- The early pioneers of science — thinkers like Newton, Galileo, and Einstein — who dared to imagine reality beyond immediate appearances, even when constrained by the paradigms of their time.
- The philosophers — known and unknown — who questioned, reflected, and wrestled with the nature of existence long before equations could formalize thought.
- The silent field of nature itself — whose spiral galaxies, ripple echoes, and living structures have patiently waited for us to see them not as accidents, but as consequences of silent intelligence.
- The independent researchers, critics, and quiet thinkers who continue to question mainstream narratives — refusing to accept borrowed knowledge without realization.
- The students of the future — who, armed with courage, humility, and true scientific spirit, will carry forward the

exploration of existence not as collectors of facts, but as discoverers of deeper consequences.

Finally, a special acknowledgment must be given to the **Structured Vacuum Energy (SVE)** — the true silent architect of all that we observe, experience, and become.

It is not just the backdrop of our studies. It is the very medium that breathes through our curiosity, balances our discoveries, and evolves our consciousness.

To all who dare to listen to the ripples beneath appearances — this work belongs as much to you as it does to me.

"We are not the creators of knowledge. We are the ripple memories — learning once again how to remember."

Sandeep Chavan

About the Author

Sandeep J Chavan is an independent researcher, physicist, and philosopher of science dedicated to reimagining the foundations of reality through consequence-driven inquiry.

Rejecting the limits of traditional academic allegiance, he has chosen the path of uncompromised exploration — seeking truth not through inherited assumptions, but through first-principle realization.

With a formal background in physics and a lifetime of deep reflection on existence, Sandeep Chavan has authored over a hundred works spanning physics, cosmology, energy dynamics, consciousness, and philosophical inquiry. His research is globally accessible through peer-acknowledged platforms, and his contributions have steadily built an alternative scientific foundation for the future.

He is the originator of the **Universal Energy Dynamics (UED)** framework — a revolutionary field-centric model that redefines space, time, mass, energy, and observation through the lens of structured vacuum intelligence.

His major published models include:

- **CMPED** (Chavan's Model of Planetary Energy Dynamics)
- **CMGED** (Chavan's Model of Galactic Energy Dynamics)
- **CEUT** (Chavan-Einstein Unified Theory)
- **SEEP** (Structured Energy Equilibrium Principle)
- And now, the culmination of these frameworks: **Chavan's Cosmic Continuum (C^3)**

Sandeep's work boldly challenges the outdated narratives of force-driven, explosion-based cosmology and instead reveals a universe alive with ripple memory, field intelligence, and evolving consciousness.

He envisions a future scientific culture where knowledge is not confined by academic orthodoxy, but expanded through consequence realization, ripple understanding, and ethical intellectual freedom.

His ongoing mission: To restore science to its true nature — not as a factory of explanations, but as a living dialogue with the universe's silent intelligence.

"Knowledge is not what we invent. It is what we remember — when we finally listen to the ripples beneath appearances."

Public Use Statement and Licensing

Public Use Statement

Chavan's Cosmic Continuum (C^3) is presented to the world as a foundational scientific contribution intended for the evolution of future inquiry, understanding, and exploration.

The ideas, models, diagrams, and conceptual frameworks contained within this report are offered in the spirit of open knowledge, collaborative advancement, and intellectual empowerment for the global scientific and philosophical community.

Readers, researchers, educators, institutions, and independent thinkers are welcome to:

- Study, reference, and cite the ideas presented here,
- Integrate the ripple-centric, consequence-driven principles into their own scientific models, philosophical inquiries, and educational materials,
- Build further theoretical, experimental, or philosophical work based on the C^3 framework, provided appropriate attribution is maintained.

The ultimate goal of this work is **not ownership**, but **expansion** — to seed a living dialogue that transcends individual authorship and serves the evolution of scientific civilization itself.

However, responsible engagement requires acknowledgment of original intellectual contribution, to maintain the integrity of the work and honor the principles of transparent knowledge sharing.

Licensing Information

Unless otherwise stated, all original ideas, models, diagrams, and written content in this report are made available under:

Creative Commons Attribution-NonCommercial 4.0 International License (CC BY-NC 4.0)

This means:

- You are free to share (copy and redistribute) and adapt (remix, transform, and build upon) the material for non-commercial purposes.
- You must give appropriate credit to **Sandeep J Chavan**, provide a link to the license, and indicate if changes were made.
- You may not use the material for commercial purposes without explicit permission from the author.

Citation Recommendation:

When citing or referencing this work, please use the following format:

Sandeep J Chavan. "Chavan's Cosmic Continuum (C^3): Manifestation and Dissolution of Universe within Structured Vacuum Energy." 2025. ORCID: 0009-0005-4190-7164.

For permissions beyond the scope of this license (e.g., commercial publications, derivative educational materials), formal inquiry must be made to the author.

Closing Note:

"Knowledge expands best when it is shared with respect, cited with integrity, and built upon with the spirit of consequence — not conquest."

Closing Note

The Universe Was Always Structured. We Are Finally Listening.

For centuries, humanity has gazed at the stars, marveled at their patterns, and constructed theories to explain what we saw. We imagined forces pushing and pulling across a vast emptiness.

We envisioned explosions birthing chaos and expansion, and we measured distances as if reality were stitched together by accidents and abstractions.

Yet beneath every orbit, every lightwave, every spiral of galaxies — the truth has been quietly, patiently waiting.

The universe was never chaos seeking order. It was — and remains — a structured, ripple-driven, memory-bearing continuum. A field so elegantly balanced that even its disturbances sing in spirals, not in noise.

We did not invent this realization. We did not impose it upon the cosmos.

We are, at last, learning to listen — to the ripples that breathe existence into form, to the memory that binds light, matter, and life, to the silent intelligence of a living field that speaks without words, without force, without violence.

Chavan's Cosmic Continuum (C^3) is not a theory trying to conquer the universe. It is a return to humility — an acceptance that reality has always been structured far beyond our earlier imaginations.

We are not discovering a new universe. We are awakening to the one that was always here.

Final Reflection:

"The universe never needed us to explain it. It only needed us to finally listen — to the spirals of memory, the ripples of consequence, and the breathing silence that creates all things."

The Defining Impact of Chavan's Cosmic Continuum (C^3)

Chavan's Cosmic Continuum (C^3) does not merely offer an alternative view of cosmology. It defines a **historic shift** — a turning point where science moves beyond patchwork theories and embraces the living, ripple-structured intelligence of existence itself.

The following five key breakthroughs **summarize the profound redefinition** C^3 brings to the foundations of physics, cosmology, and our philosophical understanding of the universe:

Consequential Formation → Not Spontaneous Creation

In C^3, the universe does not emerge from magical beginnings or miraculous explosions. Formation is a **natural consequence of ripple tension thresholds** being crossed within the Structured Vacuum Energy (SVE). Nothing happens without reason. Manifestation is not an act of creation from nothingness, but a **rebalancing event** arising intelligently from field dynamics seeking equilibrium.

Spiral Structured Evolution → Not Chaotic Explosions

Cosmic structures do not arise from debris flung outward by a random bang. Galaxies, stars, systems — all evolve through **spiral pathways** governed by ripple feedback and memory preservation. Evolution is structured, rhythmic, and spiral-driven, not chaotic or explosive. The universe breathes — it does not detonate.

Vacuum Field Dynamics → Not Force-Based Curvatures

C^3 discards the obsolete notion of forces pulling across emptiness or spacetime bending under stress. All interactions are understood as

ripple-driven corrections within a living field. The Structured Vacuum Energy (SVE) replaces the myth of inert space and force-centric explanations with **intelligent field dynamics** — simple, direct, and universal.

Energy Memory Loops → Not Singularities and Ends

Black holes are not dead ends. Singularities are illusions created by the limits of force-based thinking. In C^3, collapsed systems are **memory absorbers** — loops of ripple memory compressing into the core, preserving the record of cosmic evolution, preparing for future realignments. Nothing is lost. Everything is stored.

Cosmic Life Cycles → Not Linear Timelines

The universe is not traveling a straight line from "beginning" to "end." It moves through **eternal spiral cycles** of breathing manifestation and relaxation. Galaxies are born, evolve, dissolve, and seed new formations — **cyclically, rhythmically, endlessly** — without the need for a "start" or a "finish."

Existence is not a line. It is a **spiral of structured memory** unfolding across infinite scales.

Closing Reflection:

"C^3 is not a new idea about the universe. It is the universe — finally understood in its own language of ripples, spirals, and living memory."

Corporate Conclusion

This report — **Chavan's Cosmic Continuum** (C^3) — is **not just another theory** to be compared, debated, or archived among countless speculative models of the cosmos.

It is a **Cosmic Reorganization Charter**.

It marks the formal declaration that the era of randomness, explosions, speculative dark substances, and force-based patchwork is coming to an end.

C^3 is a structured call to the scientific civilization:

- To reset cosmology from the illusion of chaotic beginnings toward the living intelligence of structured ripple consequences.
- To replace the mythology of singular bangs with the reality of **cosmic breathing** — spiral manifestations evolving from tension, memory, and balance.
- To recognize that the universe is not a mechanical accident, but a dynamic continuum of structured intelligence — living, learning, evolving across cycles.
- To realign physics itself — from force-centric abstraction toward **field-centric realization**.

In C^3:

- Redshift is no longer proof of cosmic recession; it is ripple detuning.
- Gravitational waves are not spacetime rips; they are ripple feedback corrections.
- Black holes are not singularities; they are memory cores.
- Expansion is not the truth; cyclic breathing is.

The consequence of adopting C^3 is nothing less than a **revolution in scientific thought**:

- A new generation of theories built on consequence, not conjecture.
- A new way of teaching cosmology — integrating field intelligence at every scale.
- A new design of instruments, experiments, and cosmic mapping — tuned to ripple relics, memory fields, and spiral breathing cycles.

C^3 sets the new baseline: From randomness to ripple intelligence. From explosions to evolution. From a mechanical death march to an eternal spiral of cosmic remembrance.

This is the universe's true story — And now, we are finally ready to live, think, and evolve within it.

"The future of cosmology is not bigger telescopes chasing ancient echoes — it is deeper intelligence listening to the ripples still breathing around us."

Disclaimer

This report — **Chavan's Cosmic Continuum** (C^3) — is an original theoretical and conceptual framework developed through independent research, reflection, and scientific inquiry by **Sandeep J Chavan**.

The ideas, models, interpretations, diagrams, and terminology presented within this document are protected under applicable intellectual property rights, including copyright and citation rights, as outlined in the Public Use Statement and Licensing section.

Scientific Nature of Content:

- This report presents a field-based, ripple-consequence-driven reorganization of cosmology.
- It challenges conventional models including the Big Bang, General Relativity, and quantum force frameworks based on structured theoretical reasoning, observational motivations, and logical consequence analysis.
- The C^3 model is offered as a new lens for scientific exploration, not as a final or absolute truth.
- Readers, educators, researchers, and institutions are encouraged to critically assess, test, expand upon, or constructively challenge the ideas herein, in alignment with the principles of open, ethical scientific progress.

Non-Commercial Use:

- Redistribution, referencing, and educational use are permitted under Creative Commons Attribution-NonCommercial 4.0 International License (CC BY-NC 4.0), provided appropriate attribution is given and the material is not used for commercial purposes without formal permission.

Limits of Responsibility:

- While the C^3 model is constructed on rigorous first-principle reasoning and field-structured logic, its application, interpretation, or adaptation by third parties remains their sole responsibility.
- The author assumes no liability for misinterpretation, misapplication, or derivative works that extend beyond the intended scientific and philosophical context of this report.

Philosophical and Scientific Freedom Clause:

- This report is an exercise in scientific freedom and the spirit of responsible cosmological reimagination.
- It does not claim superiority over all existing frameworks; rather, it invites a respectful and courageous evolution of thought toward ripple-structured, consequence-regulated cosmology.

"Reality does not ask for permission to evolve. Science must listen carefully — and evolve with it."

Don't miss out!

Visit the website below and you can sign up to receive emails whenever Sandeep Chavan publishes a new book. There's no charge and no obligation.

https://books2read.com/r/B-A-EPGPC-HNNGG

BOOKS 2 READ

Connecting independent readers to independent writers.

Did you love *Chavan's Cosmic Continuum* (C^3)? Then you should read *Philosophy 2.0: A Structural Manifesto*[1] by Sandeep Chavan!

[2]

Philosophy 2.0: A Structural Manifesto by Sandeep Chavan

What if philosophy wasn't abstract theory, but the **operating manual of reality itself**?What if truth wasn't something to believe in—but something you could *align with*, feel, and live through every ripple of consequence?

Welcome to *Philosophy 2.0*—a bold and necessary redefinition of what it means to think, feel, act, and exist with structural clarity.

In this powerful manifesto, independent philosopher and systems researcher **Sandeep Chavan** offers a sweeping yet grounded framework that replaces outdated philosophical loops with a consequence-aware, energetically aligned model for modern life. Drawing from decades of

1. https://books2read.com/u/mZ6MQB

2. https://books2read.com/u/mZ6MQB

cross-disciplinary work—spanning physics, psychology, consciousness studies, ethics, education, and systems thinking—this work offers **a new foundation for intelligence itself.**

Rejecting both ideological extremes and institutional stagnation, Chavan introduces key structural principles, including:

Universal Energy Dynamics (UED): A new lens on matter, mind, and memory**The Law of Consequences (LOC)**: A unifying principle beyond morality or karma**The Mind–Soul–Consciousness Triad**: A clear and practical model of human experience**A structural rewrite of truth, justice, failure, fear, morality, purpose, and liberation**

Philosophy 2.0 also dismantles dangerous modern myths—like blind belief in science without consequence, education without alignment, and intelligence without ethics.

This book is not a doctrine.It is not a summary of others' thinking.It is a **living declaration** of a universal framework—designed not for debate, but for implementation.

Whether you're a thinker, teacher, scientist, policymaker, seeker, or student of life, this book will challenge your assumptions, refine your language, and reset your compass.

"You're not here to follow a philosophy.You're here to live one—through intelligent alignment with what your actions create."

Philosophy 2.0 is the field guide for those ready to exit illusion and enter clarity.Not a belief system.Not a reaction.A ripple-aware way of being.

Read more at https://www.sandeepjchavan.com/.

Also by Sandeep Chavan

Chavan's UED research

CMGED Research Report

Chavan's Cosmic Continuum (C^3)

Frequency

Ripples of Reality

Reality Realigned

The Great Realignment

I – The Cosmic Leverage

Echoes of War: The Russia-Ukraine Conflict and Its Global Ripples

Shattered Lives

The War Economy

Borders Redrawn

Arms and Allies

Collateral Damage

Justice in the Ashes

Somewhere in the Darkness of the Cosmos

The Blueprint of the Universe

The Cosmic Breath

The Eternal Sage Chronicles

The Sage and the Sovereign

The Family Pillar

The People's Flame

The Warrior's Soul

The Bridging Mentor

The Guiding Light

The Exodus

The Last Cycle: 2075

The Fundamental Theory of Life & Existence

Eṇergy in Action: The Soul Demystified

The Active Mind

Energy in Connection

Death: The Disconnection

Subconsciousness: The Chosen Disconnection

Desires: The Engine of Life

The Restoration of Human Knowledge

Restoring Bhagavad Gita

The War of Thoughts

Cracks in the Known
The Human Game
The Divine Dilemma
The Original Intent
The Answers

Ultra Modern Life Philosophy
The Duality of Laughter and Tears
Take It Easy
Compare to Accept

Standalone
It's Not AI, It's AHI - Amplified Human Intelligence
The Decision Paradox
The Four Sapiens
Malicious Script of Indian Polity
The IIT Legacy & Global Impact
The IIT Legacy & Global Impact
You are just a Version of Your Original
Win the Game You Didn't Choose
Always Keep Your Bags Packed
Reveal to Shield - Learn the Game of Facades
Engage Beyond Elections
Culture, Identity & Change: The Evolution of Indian Society
The Triangular Dynamics
Never Trust Without Doubt
If You Can't Change the Country, Leave the Country
Live Fearless with Fear
India: The Investment Magnet
Realignment

Powering a Billion Dreams
Universal Skill
Extension: From Brush to Algorithm
Human Side of Einstein
The Future Knows No Lines
Let's Bring the Generations to the Table
Unified Energy Dynamics
A Beat That Never Dies
Strategic Horizons
Right From Scratch
The Forgotten Innovators
Physics Fixing
The Savior Illusion
The Selfish Side
The Age of the Weakest Sapiens
Neural Wars
The Absorption: There Is No Death
Strange Therapist
Bubbles: Become a Drop
Beyond the Blue
Half-Scholar
The Law of Truth
Failure
Philosophy 2.0: A Structural Manifesto
Chaos: A Gentle Guide to the Unresolvable
Self-Help: New Syllabus

Watch for more at https://www.sandeepjchavan.com/.

www.ingramcontent.com/pod-product-compliance
Lightning Source LLC
LaVergne TN
LVHW090412210926
848480LV00002B/544